LONDON MATHEMATICAL SOCIETY LECTURE NOTE SERIES

Managing Editor: Professor J.W.S. Cassels, Department of Pure Mathematics and Mathematical Statistics, University of Cambridge, 16 Mill Lane, Cambridge CB2 1SB, England

The titles below are available from booksellers, or, in case of difficulty, from Cambridge University Press.

34 Representation theory of Lie groups, M.F. ATIYAH *et al*
36 Homological group theory, C.T.C. WALL (ed)
39 Affine sets and affine groups, D.G. NORTHCOTT
46 p-adic analysis: a short course on recent work, N. KOBLITZ
50 Commutator calculus and groups of homotopy classes, H.J. BAUES
59 Applicable differential geometry, M. CRAMPIN & F.A.E. PIRANI
66 Several complex variables and complex manifolds II, M.J. FIELD
69 Representation theory, I.M. GELFAND *et al*
76 Spectral theory of linear differential operators and comparison algebras, H.O. CORDES
77 Isolated singular points on complete intersections, E.J.N. LOOIJENGA
83 Homogeneous structures on Riemannian manifolds, F. TRICERRI & L. VANHECKE
86 Topological topics, I.M. JAMES (ed)
87 Surveys in set theory, A.R.D. MATHIAS (ed)
88 FPF ring theory, C. FAITH & S. PAGE
89 An F-space sampler, N.J. KALTON, N.T. PECK & J.W. ROBERTS
90 Polytopes and symmetry, S.A. ROBERTSON
92 Representation of rings over skew fields, A.H. SCHOFIELD
93 Aspects of topology, I.M. JAMES & E.H. KRONHEIMER (eds)
94 Representations of general linear groups, G.D. JAMES
95 Low-dimensional topology 1982, R.A. FENN (ed)
96 Diophantine equations over function fields, R.C. MASON
97 Varieties of constructive mathematics, D.S. BRIDGES & F. RICHMAN
98 Localization in Noetherian rings, A.V. JATEGAONKAR
99 Methods of differential geometry in algebraic topology, M. KAROUBI & C. LERUSTE
100 Stopping time techniques for analysts and probabilists, L. EGGHE
101 Groups and geometry, ROGER C. LYNDON
103 Surveys in combinatorics 1985, I. ANDERSON (ed)
104 Elliptic structures on 3-manifolds, C.B. THOMAS
105 A local spectral theory for closed operators, I. ERDELYI & WANG SHENGWANG
106 Syzygies, E.G. EVANS & P. GRIFFITH
107 Compactification of Siegel moduli schemes, C-L. CHAI
108 Some topics in graph theory, H.P. YAP
109 Diophantine analysis, J. LOXTON & A. VAN DER POORTEN (eds)
110 An introduction to surreal numbers, H. GONSHOR
113 Lectures on the asymptotic theory of ideals, D. REES
114 Lectures on Bochner-Riesz means, K.M. DAVIS & Y-C. CHANG
115 An introduction to independence for analysts, H.G. DALES & W.H. WOODIN
116 Representations of algebras, P.J. WEBB (ed)
117 Homotopy theory, E. REES & J.D.S. JONES (eds)
118 Skew linear groups, M. SHIRVANI & B. WEHRFRITZ
119 Triangulated categories in the representation theory of finite-dimensional algebras, D. HAPPEL
121 Proceedings of *Groups - St Andrews 1985*, E. ROBERTSON & C. CAMPBELL (eds)
122 Non-classical continuum mechanics, R.J. KNOPS & A.A. LACEY (eds)
125 Commutator theory for congruence modular varieties, R. FREESE & R. MCKENZIE
126 Van der Corput's method of exponential sums, S.W. GRAHAM & G. KOLESNIK
127 New directions in dynamical systems, T.J. BEDFORD & J.W. SWIFT (eds)
128 Descriptive set theory and the structure of sets of uniqueness, A.S. KECHRIS & A. LOUVEAU
129 The subgroup structure of the finite classical groups, P.B. KLEIDMAN & M.W. LIEBECK
130 Model theory and modules, M. PREST
131 Algebraic, extremal & metric combinatorics, M-M. DEZA, P. FRANKL & I.G. ROSENBERG (eds)
132 Whitehead groups of finite groups, ROBERT OLIVER
133 Linear algebraic monoids, MOHAN S. PUTCHA
134 Number theory and dynamical systems, M. DODSON & J. VICKERS (eds)
135 Operator algebras and applications, 1, D. EVANS & M. TAKESAKI (eds)
136 Operator algebras and applications, 2, D. EVANS & M. TAKESAKI (eds)
137 Analysis at Urbana, I, E. BERKSON, T. PECK, & J. UHL (eds)
138 Analysis at Urbana, II, E. BERKSON, T. PECK, & J. UHL (eds)
139 Advances in homotopy theory, S. SALAMON, B. STEER & W. SUTHERLAND (eds)
140 Geometric aspects of Banach spaces, E.M. PEINADOR & A. RODES (eds)
141 Surveys in combinatorics 1989, J. SIEMONS (ed)

142 The geometry of jet bundles, D.J. SAUNDERS
143 The ergodic theory of discrete groups, PETER J. NICHOLLS
144 Introduction to uniform spaces, I.M. JAMES
145 Homological questions in local algebra, JAN R. STROOKER
146 Cohen-Macaulay modules over Cohen-Macaulay rings, Y. YOSHINO
147 Continuous and discrete modules, S.H. MOHAMED & B.J. MÜLLER
148 Helices and vector bundles, A.N. RUDAKOV *et al*
149 Solitons, nonlinear evolution equations and inverse scattering, M. ABLOWITZ & P. CLARKSON
150 Geometry of low-dimensional manifolds 1, S. DONALDSON & C.B. THOMAS (eds)
151 Geometry of low-dimensional manifolds 2, S. DONALDSON & C.B. THOMAS (eds)
152 Oligomorphic permutation groups, P. CAMERON
153 L-functions and arithmetic, J. COATES & M.J. TAYLOR (eds)
154 Number theory and cryptography, J. LOXTON (ed)
155 Classification theories of polarized varieties, TAKAO FUJITA
156 Twistors in mathematics and physics, T.N. BAILEY & R.J. BASTON (eds)
157 Analytic pro-*p* groups, J.D. DIXON, M.P.F. DU SAUTOY, A. MANN & D. SEGAL
158 Geometry of Banach spaces, P.F.X. MÜLLER & W. SCHACHERMAYER (eds)
159 Groups St Andrews 1989 volume 1, C.M. CAMPBELL & E.F. ROBERTSON (eds)
160 Groups St Andrews 1989 volume 2, C.M. CAMPBELL & E.F. ROBERTSON (eds)
161 Lectures on block theory, BURKHARD KÜLSHAMMER
162 Harmonic analysis and representation theory, A. FIGA-TALAMANCA & C. NEBBIA
163 Topics in varieties of group representations, S.M. VOVSI
164 Quasi-symmetric designs, M.S. SHRIKANDE & S.S. SANE
165 Groups, combinatorics & geometry, M.W. LIEBECK & J. SAXL (eds)
166 Surveys in combinatorics, 1991, A.D. KEEDWELL (ed)
167 Stochastic analysis, M.T. BARLOW & N.H. BINGHAM (eds)
168 Representations of algebras, H. TACHIKAWA & S. BRENNER (eds)
169 Boolean function complexity, M.S. PATERSON (ed)
170 Manifolds with singularities and the Adams-Novikov spectral sequence, B. BOTVINNIK
171 Squares, A.R. RAJWADE
172 Algebraic varieties, GEORGE R. KEMPF
173 Discrete groups and geometry, W.J. HARVEY & C. MACLACHLAN (eds)
174 Lectures on mechanics, J.E. MARSDEN
175 Adams memorial symposium on algebraic topology 1, N. RAY & G. WALKER (eds)
176 Adams memorial symposium on algebraic topology 2, N. RAY & G. WALKER (eds)
177 Applications of categories in computer science, M. FOURMAN, P. JOHNSTONE, & A. PITTS (eds)
178 Lower K- and L-theory, A. RANICKI
179 Complex projective geometry, G. ELLINGSRUD *et al*
180 Lectures on ergodic theory and Pesin theory on compact manifolds, M. POLLICOTT
181 Geometric group theory I, G.A. NIBLO & M.A. ROLLER (eds)
182 Geometric group theory II, G.A. NIBLO & M.A. ROLLER (eds)
183 Shintani zeta functions, A. YUKIE
184 Arithmetical functions, W. SCHWARZ & J. SPILKER
185 Representations of solvable groups, O. MANZ & T.R. WOLF
186 Complexity: knots, colourings and counting, D.J.A. WELSH
187 Surveys in combinatorics, 1993, K. WALKER (ed)
188 Local analysis for the odd order theorem, H. BENDER & G. GLAUBERMAN
189 Locally presentable and accessible categories, J. ADAMEK & J. ROSICKY
190 Polynomial invariants of finite groups, D.J. BENSON
191 Finite geometry and combinatorics, F. DE CLERCK *et al*
192 Symplectic geometry, D. SALAMON (ed)
193 Computer algebra and differential equations, E. TOURNIER (ed)
194 Independent random variables and rearrangement invariant spaces, M. BRAVERMAN
195 Arithmetic of blowup algebras, WOLMER VASCONCELOS
196 Microlocal analysis for differential operators, A. GRIGIS & J. SJÖSTRAND
197 Two-dimensional homotopy and combinatorial group theory, C. HOG-ANGELONI, W. METZLER & A.J. SIERADSKI (eds)
198 The algebraic characterization of geometric 4-manifolds, J.A. HILLMAN
199 Invariant potential theory in the unit ball of $\mathbf{C}^n$, MANFRED STOLL
200 The Grothendieck theory of dessins d'enfant, L. SCHNEPS (ed)
201 Singularities, JEAN-PAUL BRASSELET (ed)
202 The technique of pseudodifferential operators, H.O. CORDES
203 Hochschild cohomology of von Neumann algebras, A. SINCLAIR & R. SMITH
204 Combinatorial and geometric group theory, A.J. DUNCAN, N.D. GILBERT & J. HOWIE (eds)
207 Groups of Lie type and their geometries, W.M. KANTOR & L. DI MARTINO (eds)
208 Vector bundles in algebraic geometry, N.J. HITCHIN, P. NEWSTEAD & W.M. OXBURY (eds)
210 Hilbert C*-modules, E.C. LANCE

London Mathematical Society Lecture Note Series. 210

Hilbert C*-Modules

A toolkit for operator algebraists

E.C. Lance
University of Leeds

Published by the Press Syndicate of the University of Cambridge
The Pitt Building, Trumpington Street, Cambridge CB2 1RP
40 West 20th Street, New York, NY 10011-4211, USA
10 Stamford Road, Oakleigh, Melbourne 3166, Australia

First published 1995

Library of Congress cataloguing in publication data available

British Library cataloguing in publication data available

ISBN 0 521 47910 X paperback

Transferred to digital printing 2004

Contents

	Preface	vii
1.	Modules and mappings	1
2.	Multipliers and morphisms	14
3.	Projections and unitaries	21
4.	Tensor products	31
5.	The KSGNS construction	45
6.	Stabilisation or absorption	59
7.	Full modules, Morita equivalence	69
8.	Slice maps and bialgebras	76
	Appendix: The reduced C*-algebra of a locally compact group	90
9.	Unbounded operators	94
10.	The bounded transform, unbounded multipliers	107
11.	What next?	121
	References	123
	Index	129

Preface

This book is designed as a "second course in C*-algebras". It presupposes a familiarity with the elementary theory of C*-algebras (the GNS construction, the functional calculus) such as may be found in any of the several excellent texts now available—[Dix 2], [KadRin], [Mur], [Ped], [Tak], for example. The aim, as indicated by the subtitle, is to provide the student with a collection of techniques that have shown themselves to be useful in a variety of contexts in modern C*-algebra theory.

These techniques centre round the quite elementary and natural concept of a Hilbert C*-module. As explained in Chapter 1, this is an object like a Hilbert space except that the inner product is not scalar-valued, but takes its values in a C*-algebra. The first three chapters present the elementary theory of Hilbert C*-modules and their bounded adjointable operators. From Chapter 4 onwards, tensor products figure prominently, and some knowledge of tensor products of C*-algebras (summarised at the beginning of Chapter 4) is needed.

Hilbert C*-modules have had three main areas of applications:

- the work of Rieffel and others on induced representations and Morita equivalence ([BroGreRie], [Rie 1], [Rie 2]);
- the work of Kasparov and others on KK-theory ([BaaJul], [Kas 1], [Kas 2]);
- the work of Woronowicz and others on C*-algebraic quantum group theory ([BaaSka], [Wor 5]).

There is not very much information on any of these topics in this book, since the aim is to develop a toolkit rather than to demonstrate the tools in use. However, the choice of topics has been made very much with applications in mind, and I have deliberately omitted some aspects of the theory of Hilbert C*-modules that seem to me to be internally rather than externally motivated. Broadly speaking, Chapters 4 to 6 are oriented towards KK-theory, and Chapters 8 onwards towards quantum groups, with Chapter 7 providing a short interlude on the topic of Morita equivalence. (There is, however, considerable overlap between the applications. For example, the theory of unbounded operators on Hilbert C*-modules was developed in [BaaJul] for its applications to KK-theory, and then rediscovered in [Wor 5] in connection with non-compact quantum groups.)

Readers wishing to see how Hilbert C*-modules are used in KK-theory are now well served by textbooks. As well as the overview given in [Bla] and the more technical account in [JenTho], there is a very accessible and reader-friendly account of K-theory in [Weg]. Wegge-Olsen's book, which appeared after the first draft of this book had been written, includes a useful chapter on Hilbert C*-modules. Indeed, there is some overlap (though not too much) between the earlier chapters of this book and Chapter 15 of [Weg]. Wegge-Olsen also devotes a chapter to the beautiful theorem of Cuntz, Higson and Mingo ([CunHig], [Min]) that if A is a σ-unital C*-algebra then the unitary group of $\mathcal{L}(H_A)$ is contractible. I was dissuaded from including a proof of this result only by the fact that it requires more K-theoretic preliminaries than I was prepared to deal with.

The primary goal of these notes, however, is to orient the reader towards the literature on C*-algebraic quantum groups. To that end, there is a fairly careful account in Chapter 8 of the bialgebra structure of $C_r^*(G)$, the reduced C*-algebra of a locally compact group G. Quantum groups as such do not figure in this book at all, but there is a short concluding chapter which tries to point the student towards some of the exciting current developments in this area, many of which require some knowledge of Hilbert C*-modules.

There are no exercises at the end of the chapters, but there are many places throughout the book where details (and sometimes more substantial proofs) are omitted, with or without an invitation to the reader to complete the argument. In a more general sense, the entire book (like any research-level text) should be seen as an exercise for the reader, who should approach it in a critical and participatory frame of mind, and not just as a passive recipient of knowledge.

The list of References has been kept as short as possible, maybe too short. I have cited at the end of each chapter my main (conscious) primary sources, but I may have omitted to acknowledge the work of others that I have unconsciously absorbed. To them I apologise. The restricted list of References is in line with my belief that the theory of Hilbert C*-modules should be treated as a toolkit for use in applications, rather than as an end in itself. I have, however, benefited greatly from a comprehensive unpublished bibliography on Hilbert C*-modules produced by M. Frank in 1991 (preprint, University of Leipzig). When giving references for standard results in C*-algebra theory, I have tried when possible to cite at least two of the texts mentioned in the opening paragraph above, in the hope that students will learn to consult a variety of sources and not to rely on any one book, however inspired, as their bible.

These notes are a slightly expanded form of a set of ten lectures given at a Summer School in Trondheim, August 1993. I am grateful to NorFA, the Nordic Academy for Advanced Study, for financial support, and to T. Digernes, M.B. Landstad, C. Skau and the graduate student participants in the Summer School for a most enjoyable meeting. The first draft of the book was mostly written during a sabbatical visit to the University of Pennsylvania, for which I am grateful to R.V. Kadison. I am indebted to several people (notably S. Eschmeier, M. Frank and N. Wegge-Olsen) for their helpful comments and corrections. Finally, I should like to thank Mrs A. Landford for her rapid and efficient typing of the TEXscript.

Leeds,
Dec. 1993

Chapter 1

Modules and mappings

Hilbert C*-modules first appeared in the work of Kaplansky [Kap], who used them to prove that derivations of type I AW*-algebras are inner. His idea, as explained in the introduction to [Kap], was to generalise Hilbert space by allowing the inner product to take values in a (commutative, unital) C*-algebra rather than in the field of complex numbers. Before making the formal definition, let us briefly consider why this idea might be useful.

Let A be a commutative, unital C*-algebra. By the commutative Gelfand–Naimark theorem we can identify A with $C(X)$, the algebra of continuous complex-valued functions on a compact Hausdorff space X. If X were a euclidean manifold then one would analyse it by geometric techniques, among the most important of which is the study of vector bundles over X. A vector bundle E can be described as follows. Take a fixed euclidean space H, and for each t in X let H_t be a subspace of H. Let E be the space of all continuous functions ξ from X to H such that, for all t in X, $\xi(t) \in H_t$. Then E is naturally endowed with a $C(X)$-valued inner product. Namely, if $\xi, \eta \in E$ then we define $\langle \xi, \eta \rangle$ to be the function

$$t \mapsto \langle \xi(t), \eta(t) \rangle_H .$$

Also, E has the structure of a $C(X)$-module: given ξ in E and f in $C(X)$, we define ξf to be the pointwise product $t \mapsto \xi(t)f(t)$, which is an element of E.

Since vector bundles are so effective in the study of manifolds, it is natural to want to extend the above construction to a general compact

Hausdorff space X. This time, we take H to be a Hilbert space, and for each t in X we ask that H_t should be a closed subspace of H. The construction works exactly as before, and gives rise to a $C(X)$-module E equipped with a $C(X)$-valued inner product. This is the prototypical example of a Hilbert $C(X)$-module.

Explaining his decision to work only with modules over commutative unital C*-algebras, Kaplansky wrote in a footnote to [Kap]: "The assumption of a unit element is not vital, but it seems pointless to omit it, since A will shortly be an AW*-algebra. On the other hand, extension of the theory to modules over non-commutative C*-algebras presents many difficulties." Perhaps because of this discouraging observation, there was essentially no more work on Hilbert C*-modules for a twenty-year period until the PhD thesis of Paschke [Pas]. Paschke showed that, contrary to Kaplansky's misgivings, most of the basic properties of Hilbert C*-modules are valid for modules over an arbitrary C*-algebra. At about the same time, Rieffel [Rie 1] independently developed much of the same theory, and used Hilbert C*-modules as the technical basis for his theory of induced representations of C*-algebras.

Since then the subject has grown and spread rapidly. Many of the most incisive developments were made by Kasparov, who used Hilbert C*-modules as the framework for his bivariant K-theory. Much of this book will consist of an exposition of Kasparov's work. More recently, Hilbert C*-modules have formed the technical underpinning for the C*-algebraic approach to quantum groups. The later chapters of this book are designed to orient the reader towards the literature on this subject.

We now make the formal definition of the objects that we shall be studying. Let A be a C*-algebra (not necessarily unital or commutative). An *inner-product A-module* is a linear space E which is a right A-module (with compatible scalar multiplication: $\lambda(xa) = (\lambda x)a = x(\lambda a)$ for $x \in E$, $a \in A$, $\lambda \in \mathbb{C}$), together with a map $(x,y) \mapsto \langle x,y\rangle : E \times E \to A$ such that

$$\left.\begin{array}{lll} \text{(i)} & \langle x, \alpha y + \beta z\rangle = \alpha\langle x,y\rangle + \beta\langle x,z\rangle & (x,y,z \in E,\ \alpha,\beta \in \mathbb{C}),\\ \text{(ii)} & \langle x, ya\rangle = \langle x,y\rangle a & (x,y \in E,\ a \in A),\\ \text{(iii)} & \langle y,x\rangle = \langle x,y\rangle^* & (x,y \in E),\\ \text{(iv)} & \langle x,x\rangle \geqslant 0;\quad \text{if } \langle x,x\rangle = 0 \text{ then } x = 0. & \end{array}\right\} \quad (1.1)$$

Note that condition (i) requires the inner product to be linear in its *second* variable. It follows from (iii) that the inner product is conjugate-linear in its first variable. We adopt the same convention for ordinary inner-product spaces and Hilbert spaces (so that an inner-product space is the same thing as an inner-product $\mathbb{C}$-module). This convention is in line with much of the recent research literature; but the reader should be aware that many authors use inner products that are linear in the first variable (and conjugate-linear in the second variable).

If E satisfies all the conditions for an inner-product A-module except for the second part of (iv) then we call E a *semi-inner-product A-module*. For such modules there is a useful version of the Cauchy–Schwarz inequality:

PROPOSITION 1.1. *If E is a semi-inner-product A-module and $x, y \in E$ then*

$$\langle y, x\rangle\langle x, y\rangle \leqslant \|\langle x, x\rangle\|\,\langle y, y\rangle. \tag{1.2}$$

Proof. Suppose, as we may without loss of generality, that $\|\langle x, x\rangle\| = 1$. For $a \in A$, we have

$$\begin{aligned} 0 &\leqslant \langle xa - y, xa - y\rangle \\ &= a^*\langle x, x\rangle a - \langle y, x\rangle a - a^*\langle x, y\rangle + \langle y, y\rangle \\ &\leqslant a^*a - \langle y, x\rangle a - a^*\langle x, y\rangle + \langle y, y\rangle. \end{aligned}$$

(The last line comes from the fact (1.6.8 in [Dix 2], or 1.3.5 in [Ped]) that if c is a positive element of A then $a^*ca \leqslant \|c\|a^*a$.) Now put $a = \langle x, y\rangle$ to get $a^*a \leqslant \langle y, y\rangle$, as required.

For x in E we write $\|x\| = \|\langle x, x\rangle\|^{\frac{1}{2}}$. It follows from Proposition 1.1 that $\|\langle x, y\rangle\| \leqslant \|x\|\,\|y\|$ and it is easy to deduce from this that if E is an inner-product A-module then $\|\cdot\|$ is a norm on E. If E is just a semi-inner-product A-module then $\|\cdot\|$ is a seminorm on E and, exactly as for ordinary inner-product spaces, we can construct a quotient of E that is an inner-product A-module, using Proposition 1.1. In fact, let

$$N = \{x \in E : \langle x, x\rangle = 0\}.$$

Then N is a sub-A-module of E. (It is closed under addition by Proposition 1.1.) There is a well-defined A-valued inner product on the quotient

A-module E/N given by

$$\langle x+N, y+N\rangle = \langle x, y\rangle \qquad (x, y \in E)$$

and this makes E/N into an inner-product A-module.

As well as its scalar-valued norm $\|\cdot\|$, an inner-product A-module E has an A-valued "norm" $|\cdot|$, given by $|x| = \langle x, x\rangle^{\frac{1}{2}}$. Since taking square roots of positive elements is an order-preserving operation in a C*-algebra (see 2.2.6 in [Mur] or 1.3.8 in [Ped]), it follows from Proposition 1.1 that

$$|\langle x, y\rangle| \leqslant \|x\|\,|y| \qquad (x, y \in E) \tag{1.3}$$

(where, for $a \in A$, $|a| = (a^*a)^{\frac{1}{2}}$). Notice that the norm on E makes E into a normed A-module. That is to say,

$$\|xa\| \leqslant \|x\|\,\|a\| \qquad (x \in E,\ a \in A). \tag{1.4}$$

Reason: $\langle xa, xa\rangle = a^*\langle x, x\rangle a \leqslant \|x\|^2 a^*a$. Taking square roots, we have $|xa| \leqslant \|x\|\,|a|$, from which the result follows.

The A-valued norm on an inner-product A-module is a useful device, but it needs to be handled with care. For example, it need not be the case that $|x+y| \leqslant |x| + |y|$.

An inner-product A-module which is complete with respect to its norm is called a *Hilbert A-module*, or a *Hilbert C*-module over the C*-algebra A*. Given an (incomplete) inner-product A-module E_0, one can form its completion E just as in the case of an ordinary inner-product space, and thus obtain a Hilbert A-module. This construction makes use of the completeness of the C*-algebra A: for given sequences (x_n), (y_n) in E_0 with limits x, y in E, we want to define $\langle x, y\rangle$ to be $\lim_{n\to\infty}\langle x_n, y_n\rangle$, and the completeness of A is needed to ensure that this limit exists.

Let A_0 be a pre-C*-algebra. That is to say, A_0 satisfies all the conditions to be a C*-algebra except that it need not be complete. Then A_0 has a completion A which is a C*-algebra. We can define an inner-product A_0-module in exactly the same way as an inner-product A-module, and everything that we have done so far still works, except for the completion process in the previous paragraph. In particular, if E is a Hilbert A_0-module then the inequality (1.4) enables us to extend the module action of A_0 on E by continuity to a module action of A on E and thus to make E into a Hilbert A-module.

We now want to combine the two kinds of completion that we have been considering. Suppose that A_0 is a pre-C*-algebra and that E_0 is an inner-product A_0-module. Let A, E be the completions of A_0, E_0. Then E is a module over A_0 and it can be equipped with an inner product taking values in A. We can, however, extend the module action of A_0 on E to an action of A on E so that, finally, E is a Hilbert A-module.

Let E be a Hilbert A-module and let (e_i) be an approximate unit for A. For x in E,

$$\langle x - xe_i, x - xe_i\rangle = \langle x, x\rangle - e_i\langle x, x\rangle - \langle x, x\rangle e_i + e_i\langle x, x\rangle e_i \xrightarrow{i} 0. \quad (1.5)$$

This shows that EA is dense in E. It also shows that $x1 = x$ if A has an identity 1. If A is not unital and A^+ denotes the C*-algebra obtained by adjoining an identity 1 to A then E becomes a Hilbert A^+-module if we define $x1 = x \quad (x \in X)$.

Define $\langle E, E\rangle$ to be the linear span of the set $\{\langle x, y\rangle : x, y \in E\}$. Then the closure of $\langle E, E\rangle$ is a two-sided ideal in A, call it B. If (u_i) is an approximate unit for B then the calculation in the previous paragraph shows that $xu_i \xrightarrow{i} x$ for all x in E. It follows that $E\langle E, E\rangle$ is dense in E, a fact that will have very useful consequences later. Note that B need not be the whole of A. For example, take $A = C(X)$ and let E be the Hilbert A-module described at the beginning of the chapter. If Y is a nonempty closed subset of X and $H_t = \{0\}$ whenever $t \in Y$ then it is easily seen that $B \subseteq \{f \in A : f(Y) = \{0\}\}$, which is a proper ideal of A.

It is time to have some more examples of Hilbert C*-modules. If A is a C*-algebra, then A itself is a Hilbert A-module if we define

$$\langle a, b\rangle = a^*b \qquad (a, b \in A).$$

If J is a closed right ideal in A then J is a sub-A-module of A and is therefore a Hilbert A-module.

If $\{E_i\}$ is a finite set of Hilbert A-modules then we can form the direct sum $\bigoplus E_i$. This is an A-module in the obvious way, and it becomes a Hilbert A-module if we define $\langle x, y\rangle = \sum_i \langle x_i, y_i\rangle$, where $x = (x_i)$ and $y = (y_i)$. We write E^n for the direct sum of n copies of a Hilbert A-module E.

Now let $\{E_i\}_{i\in I}$ be an infinite set of Hilbert A-modules. The construction of the direct sum in this case is more subtle, and needs to be taken carefully. We define $\bigoplus E_i$ to be the set of all sequences $x = (x_i)$, with x_i in E_i, such that $\sum_i \langle x_i, x_i\rangle$ converges in A. Note that this is a weaker condition than requiring that the series of norms $\sum_i \|\langle x_i, x_i\rangle\|$ should converge (because a series $\sum_n a_n$ of positive elements of a C*-algebra can converge without $\sum_n \|a_n\|$ converging). For $x = (x_i)$ and $y = (y_i)$ in $\bigoplus E_i$, we define $\langle x, y\rangle = \sum_i \langle x_i, y_i\rangle$. To see that this sum converges in A, note that if J is any finite subset of I then Proposition 1.1 applied to the finite direct sum $\bigoplus_J E_i$ shows that

$$\Big\|\sum_{i\in J}\langle x_i, y_i\rangle\Big\|^2 \leqslant \Big\|\sum_{i\in J}\langle x_i, x_i\rangle\Big\|\,\Big\|\sum_{i\in J}\langle y_i, y_i\rangle\Big\|.$$

Since x and y are in $\bigoplus E_i$, the right-hand side of this inequality can be made small whenever J is disjoint from some finite subset of I. This is what is needed to show that the inner product on $\bigoplus E_i$ is well-defined; and this inner product evidently makes $\bigoplus E_i$ into an inner-product A-module. In fact, it is complete, and is therefore a Hilbert A-module. (Exercise: prove this.)

If H is a Hilbert space then the algebraic (vector space) tensor product $H\otimes_{\text{alg}}A$ (which is a right A-module) has an A-valued inner product given on simple tensors by

$$\langle \xi\otimes a, \eta\otimes b\rangle = \langle \xi, \eta\rangle a^*b \qquad (\xi, \eta \in H,\ a, b \in A).$$

(To see that the inner product given in this way is positive definite, write $t = \left\langle \sum \xi_i\otimes a_i, \sum \xi_i\otimes a_i\right\rangle$, and let $\{\varepsilon_k\}$ be an orthonormal basis for the finite-dimensional subspace of H spanned by the ξ_i. If $\xi_i = \sum_k \lambda_{ik}\varepsilon_k$ then $t = \sum_k \left(\sum_i \lambda_{ik}a_i\right)^*\left(\sum_i \lambda_{ik}a_i\right)$. So $t \geqslant 0$; and if $t = 0$ then $\sum_i \lambda_{ik}a_i = 0$ for all k, from which $\sum_i \xi_i\otimes a_i = 0$.) Thus $H\otimes_{\text{alg}}A$ is an inner-product A-module, and we denote its completion by $H\otimes A$. If $\{\varepsilon_i\}$ is an orthonormal basis for H then $H\otimes A$ can be naturally identified with the Hilbert A-module $\bigoplus A_i$ defined in the previous paragraph, where each A_i is a copy of A. In the case where H is a separable, infinite-dimensional Hilbert space, the Hilbert A-module $H\otimes A$ is often denoted by H_A. It plays an important special role in the theory, as we shall see in subsequent chapters.

For a final example, suppose that A is a unital C*-algebra, B is a C*-subalgebra of A containing the identity, and $\psi: A \to B$ is a linear norm-reducing idempotent. Such a map is called a conditional expectation from A to B. A conditional expectation is always a positive map and satisfies

$$\psi(bac) = b\psi(a)c \qquad (a \in A,\ b, c \in B)$$

(see [Tak]). It is said to be faithful if

$$a \in A,\ a \geqslant 0,\ \psi(a) = 0 \implies a = 0.$$

Suppose that E is a Hilbert A-module, with inner product $\langle \cdot, \cdot \rangle_A$. Then E is a semi-inner-product B-module under the inner product given by

$$\langle x, y \rangle_B = \psi\big(\langle x, y \rangle_A\big).$$

If ψ is faithful then E is an inner-product B-module.

The above construction is called *localisation*. It will be studied further in Chapter 5 (where we shall eliminate the condition that A and B should be unital).

One would like to think that Hilbert C*-modules behave like Hilbert spaces, and in some ways they do. For example, if E is a Hilbert A-module and $x \in E$ then it is easy to check that

$$\|x\| = \sup\{\|\langle x, y \rangle\| : y \in E,\ \|y\| \leqslant 1\}$$

(a fact that we shall frequently use). But there is one fundamental way in which Hilbert C*-modules differ from Hilbert spaces. Given a closed submodule F of a Hilbert A-module E, define

$$F^\perp = \{y \in E : \langle x, y \rangle = 0 \ \ (x \in F)\}.$$

Then $F^\perp$ is also a closed submodule of E. But E is not (usually) equal to $F \oplus F^\perp$ (and $F^{\perp\perp}$ is usually larger than F). For example, take $A = C(X)$ and let Y be a nonempty closed subset of X whose complement is dense in X. Let $E = A$ and let $F = \{f \in A : f(Y) = \{0\}\}$. In this case, $F^\perp = \{0\}$.

Since the whole theory of Hilbert spaces and their operators is based on the use of orthogonal complements, it is clear that there will be obstacles to developing an analogous theory for Hilbert C*-modules. Nevertheless, it is useful to use Hilbert space ideas as a guide, adding extra conditions when necessary to obtain a theory which works for Hilbert C*-modules. With this in mind, we now introduce some important classes of operators on Hilbert C*-modules.

Suppose that E, F are Hilbert A-modules. We define $\mathcal{L}(E,F)$ to be the set of all maps $t: E \to F$ for which there is a map $t^*: F \to E$ such that

$$\langle tx, y\rangle = \langle x, t^*y\rangle \qquad (x \in E,\ y \in F).$$

It is easy to see that t must be A-linear (that is, t is linear and $t(xa) = t(x)a$ for all $x \in E$, $a \in A$). For each x in the unit ball E_1 of E, define $f_x: F \to A$ by

$$f_x(y) = \langle tx, y\rangle \qquad (y \in F).$$

Then $\|f_x(y)\| \leqslant \|t^*y\|$ for all x in E_1. It follows from the Banach–Steinhaus theorem that the set $\{\|f_x\|: x \in E_1\}$ is bounded, and this shows that the mapping t is bounded. We call $\mathcal{L}(E,F)$ the set of *adjointable* maps from E to F.

Thus every element of $\mathcal{L}(E,F)$ is a bounded A-linear map. It is important to realise that the converse is false: a bounded A-linear map need not be adjointable. For example, let X be a compact Hausdorff space and let Y be a closed nonempty subset of X with dense complement. Let $F = A = C(X)$, let $E = \{f \in A: f(Y) = \{0\}\}$ and let $i: E \to F$ be the inclusion mapping. A simple calculation shows that if i were adjointable and if 1 denotes the identity element of A then $i^*(1)$ would have to be equal to 1. But $1 \notin E$, and so i cannot be adjointable.

It is clear that if $t \in \mathcal{L}(E,F)$ then $t^* \in \mathcal{L}(F,E)$. If also G is a Hilbert A-module and $s \in \mathcal{L}(F,G)$ then $st \in \mathcal{L}(E,G)$. In particular, $\mathcal{L}(E,E)$, which we abbreviate to $\mathcal{L}(E)$, is a $*$-algebra. In fact, it is a C*-algebra. For it is a closed subset of the algebra of all bounded operators on E, and therefore a Banach algebra; and the calculation

$$\begin{aligned}\|t^*t\| &\geqslant \sup\{\|\langle t^*tx, x\rangle\| : x \in E_1\}\\ &= \sup\{\|\langle tx, tx\rangle\| : x \in E_1\} = \|t\|^2\end{aligned}$$

shows that the operator norm satisfies the C*-condition.

It will occasionally be convenient to use the notations $\mathcal{L}_A(E)$, $\mathcal{L}_A(E,F)$ in place of $\mathcal{L}(E)$, $\mathcal{L}(E,F)$, to make explicit the underlying C*-algebra. This will be especially important when, as sometimes happens, we are dealing with a space E that can be a Hilbert C*-module over more than one C*-algebra.

PROPOSITION 1.2. *If $t \in \mathcal{L}(E,F)$ and $x \in E$ then $|tx|^2 \leqslant \|t\|^2\,|x|^2$ and $|tx| \leqslant \|t\|\,|x|$.*

Proof. Let ρ be a state of A. By repeated application of the Cauchy–Schwarz inequality to the semi-inner product $\rho(\langle\,\cdot\,,\,\cdot\,\rangle)$ on E, we obtain

$$\begin{aligned}
\rho(\langle t^*tx, x\rangle) &\leqslant \rho(\langle t^*tx, t^*tx\rangle)^{\frac{1}{2}}\rho(\langle x,x\rangle)^{\frac{1}{2}} \\
&= \rho(\langle (t^*t)^2x, x\rangle)^{\frac{1}{2}}\rho(|x|^2)^{\frac{1}{2}} \\
&\leqslant \rho(\langle (t^*t)^2x, (t^*t)^2x\rangle)^{\frac{1}{4}}\rho(|x|^2)^{\frac{1}{2}+\frac{1}{4}} \\
&\ \ \vdots \\
&\leqslant \rho(\langle (t^*t)^{2^n}x, x\rangle)^{2^{-n}}\rho(|x|^2)^{\frac{1}{2}+\frac{1}{4}+\cdots+2^{-n}} \\
&\leqslant (\|x\|^2)^{2^{-n}}\|t^*t\|\,\rho(|x|^2)^{1-2^{-n}}.
\end{aligned}$$

As $n \to \infty$ we obtain $\rho(\langle tx, tx\rangle) \leqslant \|t\|^2\rho(|x|^2)$. Since this holds for all states ρ of A, we have $|tx|^2 \leqslant \|t\|^2\,|x|^2$. The second inequality follows on taking square roots.

We now introduce a class of operators analogous to the finite-rank operators on a Hilbert space. Let E, F be Hilbert A-modules. For x in E and y in F, define $\theta_{x,y}\colon F \to E$ by

$$\theta_{x,y}(z) = x\langle y, z\rangle \qquad (z \in F).$$

It is easy to check that $\theta_{x,y} \in \mathcal{L}(F,E)$, with $(\theta_{x,y})^* = \theta_{y,x}$, and also that the following relations hold (where G is a Hilbert A-module):

$$\left.\begin{aligned}
\theta_{x,y}\theta_{u,v} &= \theta_{x\langle y,u\rangle,v} = \theta_{x,v\langle u,y\rangle} && (u \in F,\ v \in G), \\
t\theta_{x,y} &= \theta_{tx,y} && (t \in \mathcal{L}(E,G)), \\
\theta_{x,y}s &= \theta_{x,s^*y} && (s \in \mathcal{L}(G,F)).
\end{aligned}\right\} \tag{1.6}$$

We denote by $\mathcal{K}(F, E)$ the closed linear subspace of $\mathcal{L}(F, E)$ spanned by $\{\theta_{x,y} : x \in E, y \in F\}$, and we write $\mathcal{K}(E)$ for $\mathcal{K}(E, E)$. It follows from the above relations that $\mathcal{K}(E)$ is an ideal (by which we always mean a closed, two-sided ideal unless otherwise indicated) in $\mathcal{L}(E)$. Elements of $\mathcal{K}(F, E)$ are often referred to as "compact" operators. But considered as operators between the Banach spaces F and E they need not be compact, so we shall avoid this terminology. (Example: Let A be a unital C*-algebra. Then $\theta_{1,1}$ is the identity operator on the Hilbert A-module A. So the identity operator is in $\mathcal{K}(A)$. But it is not a compact operator on the Banach space A, unless A is finite-dimensional.)

In particular examples, $\mathcal{K}(E)$ is sometimes quite easy to describe. For example, in the case when $E = A$ we have $\mathcal{K}(E) \cong A$, the isomorphism being given by identifying $\theta_{a,b}$ with the operation of left multiplication by ab^*. We shall frequently make use of this isomorphism, and the reader is urged to establish its existence carefully. The proof will make use of the fact that products are dense in any C*-algebra (this follows from the existence of an approximate unit). If A is unital then $\mathcal{K}(A) = \mathcal{L}(A)$, since it is easily verified that any t in $\mathcal{L}(A)$ consists of left multiplication by $t(1)$. When A is not unital, $\mathcal{L}(A)$ is in general much bigger than $\mathcal{K}(A)$, as we shall see in the next chapter.

If H is a Hilbert space and $\xi, \eta \in H$ then we denote by $\xi \cdot \eta$ the rank-one operator

$$\zeta \mapsto \xi\langle\eta, \zeta\rangle \qquad (\zeta \in H).$$

For the Hilbert A-module $H \otimes A$, we have $\mathcal{K}(H \otimes A) \cong \mathcal{K}(H) \otimes A$, where $\mathcal{K}(H)$ is the C*-algebra of compact operators on H and $\mathcal{K}(H) \otimes A$ denotes the C*-algebraic tensor product of $\mathcal{K}(H)$ and A. The algebraic component of this assertion is easily verified: the correspondence is given by identifying $\theta_{\xi \otimes a, \eta \otimes b}$ with $(\xi \cdot \eta) \otimes ab^*$ where, as before, ab^* stands for left multiplication by ab^*. (Writing c for ab^*, we note for future use that $\mathcal{K}(H \otimes A)$ is generated by elements of the form $(\xi \cdot \eta) \otimes c$.) The analytic component of the proof, in which the norm on $\mathcal{K}(H \otimes A)$ is identified with the C*-tensor product norm on $\mathcal{K}(H) \otimes A$, is deferred until Chapter 4, where we shall study tensor products systematically.

If E, F are Hilbert A-modules then $\mathcal{K}(E^m, F^n)$ can be identified with the set of $m \times n$ matrices over $\mathcal{K}(E, F)$. If

$$x = \begin{pmatrix} x_1 \\ \vdots \\ x_m \end{pmatrix} \in E^m, \quad y = \begin{pmatrix} y_1 \\ \vdots \\ y_n \end{pmatrix} \in F^n,$$

then $\theta_{x,y}$ corresponds to the matrix with entries θ_{x_i,y_j}. We can also identify $\mathcal{L}(E^m, F^n)$ with the set of $m \times n$ matrices over $\mathcal{L}(E,F)$. To see this, let p_i denote the projection from E^m onto its ith coordinate $(1 \leqslant i \leqslant m)$. Then p_i is adjointable, and p_i^* is the inclusion mapping of E as the ith coordinate of E^m. Similarly let $q_j \in \mathcal{L}(F^n, F)$ be the projection onto the jth coordinate $(1 \leqslant j \leqslant n)$. Given t in $\mathcal{L}(E^m, F^n)$, we associate with t the $m \times n$ matrix over $\mathcal{L}(E,F)$ whose (i,j)-entry is $q_j t p_i^*$.

For operators on Hilbert spaces, it is usual to use many topologies other than the norm topology. In the analysis of operators on C*-modules, we shall only need one such topology. This should probably be called the strong* topology, but it is usually called the strict topology and we shall follow this usage. The strict topology on $\mathcal{L}(E,F)$ is defined to be the topology given by the seminorms

$$t \mapsto \|tx\| \quad (x \in E), \qquad t \mapsto \|t^*y\| \quad (y \in F).$$

PROPOSITION 1.3. *If E, F are Hilbert A-modules then the unit ball of $\mathcal{K}(E,F)$ is strictly dense in the unit ball of $\mathcal{L}(E,F)$.*

Proof. Let (e_i) be an approximate unit for the C*-algebra $\mathcal{K}(E)$. We shall show that $e_i x \xrightarrow{i} x$ for all x in E. It then easily follows that, for any t in $\mathcal{L}(E,F)$,

$$\|(te_i - t)x\| \xrightarrow{i} 0 \quad (x \in E), \qquad \|(e_i t^* - t^*)y\| \xrightarrow{i} 0 \quad (y \in F).$$

That is, (te_i) converges strictly to t. Since $\mathcal{K}(E)$ is an ideal in $\mathcal{L}(E)$, this will complete the proof.

We have already seen that $E\langle E,E\rangle$ is dense in E. Thus (since $\{e_i\}$ is bounded) it will suffice to prove that $e_i x \xrightarrow{i} x$ when $x \in E\langle E,E\rangle$. But if $x = w\langle u,v\rangle$ then

$$e_i x = e_i \theta_{w,u}(v) \xrightarrow{i} \theta_{w,u}(v) = x,$$

as required.

We shall require the following result at just one, crucial, point, in the Appendix to Chapter 8.

PROPOSITION 1.4. *Let E be a Hilbert A-module, and let B be a strictly dense C^*-subalgebra of $\mathcal{L}(E)$. Then the unit ball of B is strictly dense in the unit ball of $\mathcal{L}(E)$.*

Proof. You can take your favourite proof of the Kaplansky density theorem and it will work more or less word for word in this context. So we shall just give a brief outline of the proof, following the argument in [Dix 1], §I.3.5.

The function $f: \mathbf{R} \to [-1,1]$ given by $f(\lambda) = 2\lambda(1+\lambda^2)^{-1}$ maps $[-1,1]$ bijectively onto itself, and so f has a right inverse g (which is in fact given by $g(\mu) = \mu^{-1}(1-\sqrt{1-\mu^2})$ if $0 < |\mu| \leqslant 1$, $g(0) = 0$). If $s, t \in \mathcal{L}(E)_{\text{sa}}$ (the set of selfadjoint elements in $\mathcal{L}(E)$) then we have the identity

$$f(s) - f(t) = 2(1+s^2)^{-1}(s-t)(1+t^2)^{-1} - \tfrac{1}{2}f(s)(s-t)f(t),$$

from which it follows that the map $t \mapsto f(t)$ is strictly continuous from $\mathcal{L}(E)_{\text{sa}}$ to the unit ball of $\mathcal{L}(E)_{\text{sa}}$.

Let $t \in \mathcal{L}(E)_{\text{sa}}$, with $\|t\| \leqslant 1$, and let $s = g(t)$. We can find a directed net (s_i) of selfadjoint elements of B with $s_i \to s$ strictly. Therefore $f(s_i) \to f(s) = t$ strictly, and we conclude that t is in the strict closure of the unit ball of B.

For a general t in the unit ball of $\mathcal{L}(E)$, let $T = \left(\begin{smallmatrix} 0 & t \\ t^* & 0 \end{smallmatrix}\right)$ in the unit ball of $M_2(\mathcal{L}(E)) \cong \mathcal{L}(E \oplus E)$. Then T is selfadjoint, and by the first part of the proof we can find a directed net (T_i) in the unit ball of $M_2(B)$ with $T_i \to T$ strictly. If $T_i = \left(\begin{smallmatrix} * & t_i \\ * & * \end{smallmatrix}\right)$ then $t_i \to t$ strictly in the unit ball of $\mathcal{L}(E)$.

The enquiring reader, who took seriously the idea that the theory of Hilbert C*-modules should be modelled on that of Hilbert spaces, will no doubt already have asked the question: What about the Riesz–Fréchet theorem? Are bounded linear functionals given by inner products? If this question is formulated in the right way then it has a positive answer, as follows.

Let E be a Hilbert A-module, fix x in E and let

$$t_x y = \langle x, y \rangle \qquad (y \in E). \tag{1.7}$$

Then t_x is a bounded A-linear mapping from E to A. Furthermore, the calculation

$$\langle t_x y, a\rangle = \langle x, y\rangle^* a = \langle y, xa\rangle \qquad (x, y \in E,\ a \in A)$$

shows that t_x is adjointable, with $t_x^*(a) = xa$. Hence $t_x \in \mathcal{L}(E, A)$. Slightly less obviously, $t_x \in \mathcal{K}(E, A)$. The reason for this is that

$$\theta_{a,z}(y) = a\langle z, y\rangle = \langle za^*, y\rangle = t_{za^*}(y) \qquad (y, z \in E,\ a \in A). \tag{1.8}$$

Thus $t_x \in \mathcal{K}(E, A)$ whenever x is of the form $x = za^*$. Since EA is dense in E, and it is easily verified that the map $x \mapsto t_x\colon E \to \mathcal{L}(E, A)$ is isometric, it follows that $t_x \in \mathcal{K}(E, A)$ for all x in E. Thus the space $\mathcal{K}(E, A)$ is the natural generalisation to the Hilbert C*-module setting of the space of bounded linear functionals on a Hilbert space. The appropriate form of the Riesz–Fréchet theorem for Hilbert C*-modules should therefore be the assertation that every element of $\mathcal{K}(E, A)$ is given by an inner product as in (1.7), or in other words is of the form t_x for some x in E. This assertion is true, and its proof is left as an exercise (using (1.8)).

If A is unital then every element of $\mathcal{L}(E, A)$ is given by an inner product (and therefore $\mathcal{K}(E, A) = \mathcal{L}(E, A)$). In fact, it is trivially verified that if $t \in \mathcal{L}(E, A)$ then $t = t_x$ where $x = t^*(1)$. If A is nonunital then in general $\mathcal{L}(E, A) \neq \mathcal{K}(E, A)$, and so elements of $\mathcal{L}(E, A)$ may not be given by inner products.

There has been a fair amount of published work on the topic of so-called "self-dual" Hilbert C*-modules, that is to say, Hilbert A-modules E for which every bounded A-linear map from E to A (not assumed adjointable) is given by an inner product. (See for example [Pas], [Fil], [Fra].) We shall have no use for such results, so we shall not pursue this topic. For that matter, we shall have no occasion to use the above Riesz–Fréchet theorem, which is included as a curiosity and an excuse for an exercise, rather than as a useful result.

References for Chapter 1: [Pas], [Rie 1], [Bus], [Kas 1], [JenTho].

Chapter 2

Multipliers and morphisms

If A is a nonunital C*-algebra then there are various ways of embedding A in a unital algebra. The simplest method is to adjoin an identity to A. The resulting C*-algebra A^+ contains A as an ideal of codimension one. This is clearly the minimal unital algebra containing A. At the other extreme, we could try to find a maximal C*-algebra B containing A as an ideal. It is necessary to impose some restriction on B, otherwise no such algebra will exist. Indeed, if A is an ideal in B, and C is an arbitrary C*-algebra, then A (identified with $A \oplus 0$) is an ideal in the larger algebra $B \oplus C$. Note that $B \oplus C$ contains an ideal C (more precisely $0 \oplus C$) that has zero intersection with A. This leads us to make the following definition: if A is an ideal in B then we call A an *essential* ideal if there is no nonzero ideal of B that has zero intersection with A. An equivalent condition is

$$b \in B,\ bA = \{0\} \Longrightarrow b = 0. \tag{2.1}$$

We shall show that for any C*-algebra A there is a unique (up to isomorphism) C*-algebra which contains A as an essential ideal and is maximal in the sense that any other such algebra can be embedded in it. This algebra is called the multiplier algebra of A and is denoted by $M(A)$. It is easy to see that if A is unital then $M(A)$ must be equal to A, so the interest lies in the case where A is nonunital.

Let us pause to consider what happens in the commutative case. If A is a nonunital commutative C*-algebra then we can identify A with the algebra $C_0(X)$ of continuous functions vanishing at infinity on the locally compact

Hausdorff space X. Embedding A as an ideal in a (commutative, unital) C*-algebra B corresponds to embedding X as an open subspace of a compact space Y. The embedding $f \mapsto \hat{f}: C_0(X) \to C(Y)$ is given by setting $\hat{f} = f$ on X and $\hat{f} = 0$ off X. It is easy to see that A is essential in B if and only if X is dense in Y. The minimal choice for Y is the one-point compactification of X, and in this case $B = A^+$. At the other extreme, there is a maximal choice for Y, namely the Stone–Čech compactification βX of X. So in the commutative case we can identify the multiplier algebra $M(C_0(X))$ with $C(\beta X)$. Thus what follows can be regarded as a noncommutative analogue of the Stone–Čech compactification.

Traditionally, multiplier algebras of C*-algebras have been constructed through the use of double centralisers, and this is the way that the topic is handled in [Mur] and [Ped]. We shall avoid double centralisers altogether and instead approach multipliers via Hilbert C*-modules. Given a C*-algebra A, we consider A as a Hilbert A-module. As shown in the previous chapter, we can identify A with $\mathcal{K}(A)$. It will turn out that $M(A)$ can be identified with $\mathcal{L}(A)$. Indeed, we already know that $\mathcal{K}(A)$ is an ideal in $\mathcal{L}(A)$, and it is easily seen using (2.1) that it is an essential ideal. To complete the identification of $\mathcal{L}(A)$ with $M(A)$, we need to investigate representations of C*-algebras over Hilbert C*-modules.

Let C be a C*-algebra and let E be a Hilbert C-module. We say that a $*$-homomorphism $\alpha: A \to \mathcal{L}(E)$ is *nondegenerate* if $\alpha(A)E$ is dense in E. (When writing a product of sets, we always mean the *linear span* of products of elements of the sets, so that $\alpha(A)E$ by definition means

$$\Big\{\sum_{i=1}^{n} \alpha(a_i)x_i : n \geqslant 1,\ a_1,\dots,a_n \in A,\ x_1,\dots,x_n \in E\Big\}.$$

This convention will be in force throughout the remainder of the book whenever we write the product of two sets by juxtaposition.) Note that the proof of Proposition 1.3 amounted to showing that the inclusion map $\mathcal{K}(E) \subseteq \mathcal{L}(E)$ is nondegenerate.

PROPOSITION 2.1. *Let A, B, C be C*-algebras, such that A is an ideal in B, and let E be a Hilbert C-module. Suppose that $\alpha: A \to \mathcal{L}(E)$ is a nondegenerate $*$-homomorphism. Then α extends uniquely to a $*$-homomorphism $\bar{\alpha}: B \to \mathcal{L}(E)$. If α is injective and A is essential in B then $\bar{\alpha}$ is injective.*

Proof. Let (e_j) be an approximate unit for A. For b in B, $a_1, \ldots, a_n$ in A and $\xi_1, \ldots, \xi_n$ in E, we have

$$\begin{aligned} \Big\|\sum_{i=1}^{n} \alpha(ba_i)\xi_i\Big\| &= \lim_j \Big\|\sum_i \alpha(be_j a_i)\xi_i\Big\| \\ &= \lim_j \Big\|\alpha(be_j)\sum_i \alpha(a_i)\xi_i\Big\| \\ &\leqslant \|b\| \Big\|\sum_i \alpha(a_i)\xi_i\Big\|. \end{aligned}$$

Thus the map $\sum_i \alpha(a_i)\xi_i \mapsto \sum_i \alpha(ba_i)\xi_i$ is well-defined and continuous, and since α is nondegenerate it extends by continuity to a bounded map $\bar{\alpha}(b)$ on E. A similar manipulation, using (e_j), shows that $\bar{\alpha}(b^*)$ acts as an adjoint for $\bar{\alpha}(b)$, and so $\bar{\alpha}(b) \in \mathcal{L}(E)$. It is easy to verify that $\bar{\alpha}$ is a $*$-homomorphism. The uniqueness is an obvious consequence of the nondegeneracy of α, since $\bar{\alpha}(b)\sum_i \alpha(a_i)\xi_i$ must be equal to $\sum_i \alpha(ba_i)\xi_i$. Finally, if α is injective then $\ker(\bar{\alpha})$ is an ideal in B which has zero intersection with A. If A is essential then this ideal must be zero.

Applying Proposition 2.1 in the case where $C = E = A$ and α is the canonical embedding of A in $\mathcal{L}(A)$ (that is, α consists of the identification of A with $\mathcal{K}(A)$), we see that any C*-algebra B that contains A as an essential ideal embeds in $\mathcal{L}(A)$. This shows that $\mathcal{L}(A)$ has the maximality property required of $M(A)$. The other property required of $M(A)$ is its uniqueness, and this can also be deduced from Proposition 2.1, as follows.

Suppose that B is a maximal essential extension of A. (What we mean by this is that A is an essential ideal in B, and that if A is also an essential ideal in another C*-algebra C then the identity map on A extends to an embedding $\beta: C \to B$.) Since A is an essential ideal in $\mathcal{L}(A)$, it follows that there is an injection $\beta: \mathcal{L}(A) \to B$ whose restriction to A is the identity map. By Proposition 2.1, the canonical embedding $\alpha: A \to \mathcal{L}(A)$ has an injective extension $\bar{\alpha}: B \to \mathcal{L}(A)$. By Proposition 2.1 again (applied this time to A as an essential ideal in $\mathcal{L}(A)$ rather than as an essential ideal in B), α has a unique extension to a $*$-homomorphism from $\mathcal{L}(A)$ to $\mathcal{L}(A)$. But the identity map on $\mathcal{L}(A)$ is one such extension; and $\bar{\alpha}\beta$ is another. So $\bar{\alpha}\beta$ must be the identity on $\mathcal{L}(A)$ and therefore $\bar{\alpha}$ is surjective. Thus $\bar{\alpha}$ is a $*$-isomorphism between B and $\mathcal{L}(A)$.

That completes the proof that $\mathcal{L}(A)$ is, up to isomorphism, the unique maximal essential extension of A. We summarise this result in the following theorem. Thereafter, we shall write $M(A)$ for $\mathcal{L}(A)$, calling it the multiplier algebra of A and identifying A with $\mathcal{K}(A)$.

THEOREM 2.2. *Let A be a C*-algebra.*

(i) *The algebra $\mathcal{L}(A)$ is an essential extension of $\mathcal{K}(A)$ which is maximal in the sense that if $\mathcal{K}(A)$ is an essential ideal in a C*-algebra C then there is an injective $*$-homomorphism from C to $\mathcal{L}(A)$ whose restriction to $\mathcal{K}(A)$ is the identity map.*

(ii) *If the C*-algebra B is a maximal essential extension of A then there is a $*$-isomorphism from B onto $\mathcal{L}(A)$ whose restriction to A is the canonical map from A to $\mathcal{K}(A)$.*

There is a useful result that often enables one to find the multiplier algebra of a given C*-algebra A. This is the following proposition, which says that if you can represent A faithfully and nondegenerately as an algebra of adjointable operators on some Hilbert C*-module then you can identify $M(A)$ with the idealiser of (the image of) A.

PROPOSITION 2.3. *Let A, C be C*-algebras and let E be a Hilbert C-module. Suppose that $\alpha: A \to \mathcal{L}(E)$ is a nondegenerate injective $*$-homomorphism and let B be the idealiser of $\alpha(A)$ in $\mathcal{L}(E)$; that is,*

$$B = \{s \in \mathcal{L}(E): s\alpha(A) \subseteq \alpha(A) \text{ and } \alpha(A)s \subseteq \alpha(A)\}.$$

Then α extends to a $*$-isomorphism between $M(A)$ and B.

Proof. It is clear from the definition of B that $\alpha(A)$ is an ideal in B. In fact, $\alpha(A)$ is essential in B: for if $b \in B$ and $b\alpha(A) = \{0\}$ then $b\alpha(A)E = \{0\}$, so $bE = \{0\}$ (because α is nondegenerate) and therefore $b = 0$. So by (2.1) $\alpha(A)$ is essential. By Theorem 2.2, the proof will be complete if we show that B is a maximal essential extension of $\alpha(A)$.

Suppose then that A is an essential ideal in a C*-algebra D. By Proposition 2.1, the map $\alpha: A \to \mathcal{L}(E)$ extends to an injective $*$-homomorphism $\bar{\alpha}: D \to \mathcal{L}(E)$. Since A is an ideal in D it is evident that $\bar{\alpha}(D) \subseteq B$. Thus D embeds in B via a mapping that extends α. This shows that B has the required maximality property.

As an illustration of the power of Proposition 2.3, we use it to give a one-sentence proof of a substantial result of Kasparov:

THEOREM 2.4. *If E is a Hilbert A-module then $\mathcal{L}(E) \cong M(\mathcal{K}(E))$.*

Proof. The inclusion map $i: \mathcal{K}(E) \to \mathcal{L}(E)$ is nondegenerate, and the idealiser of $\mathcal{K}(E)$ is $\mathcal{L}(E)$, so by Proposition 2.3 i extends to a $*$-isomorphism between $M(\mathcal{K}(E))$ and $\mathcal{L}(E)$.

Putting $A = \mathbf{C}$ in Theorem 2.4, we see that if H is a Hilbert space then $M(\mathcal{K}(H)) = \mathcal{L}(H)$.

As another illustration of the use of Proposition 2.3, we show that, for a locally compact Hausdorff space X, $M(C_0(X))$ can be identified with the algebra $C_b(X)$ of continuous bounded functions on X (in accordance with our previous expectation that $M(C_0(X))$ should be equal to $C(\beta X)$). To see this, represent $C_0(X)$ as multiplication operators on the Hilbert space $l^2(X)$. It is not hard to see that its idealiser in $\mathcal{L}(l^2(X))$ consists of (multiplication by) continuous bounded functions on X.

Let A, B be C*-algebras. We define a *morphism* from A to B to be a nondegenerate $*$-homomorphism from A to $M(B)$, and we denote the set of all morphisms from A to B by $\mathrm{Mor}(A, B)$. (Recall that we are identifying $M(B)$ with $\mathcal{L}(B)$, so that nondegeneracy for $\alpha \in \mathrm{Mor}(A, B)$ means that $\alpha(A)B$ is dense in B.) By Proposition 2.1, if $\alpha \in \mathrm{Mor}(A, B)$ then α extends uniquely to a $*$-homomorphism $\bar{\alpha}: M(A) \to M(B)$, and if α is injective then so is $\bar{\alpha}$. It turns out that $\mathrm{Mor}(A, B)$, rather than the apparently more natural class of $*$-homomorphisms from A to B, forms the most useful category of multiplicative mappings between C*-algebras. There is a big gain in flexibility by allowing maps that take values in $M(B)$ rather than in B itself. To pay for this, it is necessary to impose the condition of nondegeneracy. This causes some technical complications, but these can be handled through the next proposition, in which we use the strict topology defined at the end of the previous chapter.

PROPOSITION 2.5. *Let A, B be C*-algebras and let E be a Hilbert B-module. For a $*$-homomorphism $\alpha: A \to \mathcal{L}(E)$, the following conditions are equivalent:*

(i) *α is nondegenerate;*

(ii) *α is the restriction to A of a unital $*$-homomorphism $\bar{\alpha}$ from $M(A)$ to $\mathcal{L}(E)$ which is strictly continuous on the unit ball;*

(iii) *for some approximate unit (e_i) of A, $\alpha(e_i) \to 1$ strictly, where 1 is the identity map on E.*

Proof. (i) $\Rightarrow$ (ii): Suppose that $x_i \to x$ strictly in the unit ball of $M(A)$, so that $\|(x_i - x)a\| \xrightarrow{i} 0$ for all a in A. For a in A and ξ in E we have

$$\|(\bar{\alpha}(x_i) - \bar{\alpha}(x))\alpha(a)\xi\| = \|\alpha((x_i - x)a)\xi\| \xrightarrow{i} 0,$$

where $\bar{\alpha}: M(A) \to \mathcal{L}(E)$ is the extension of α given by Proposition 2.1. Since α is nondegenerate and $\|x_i - x\|$ is bounded by 2, it follows that

$$\|(\bar{\alpha}(x_i) - \bar{\alpha}(x))\zeta\| \xrightarrow{i} 0$$

for all ζ in E. The same applies with x_i^*, x^* in place of x_i, x. This shows that $\bar{\alpha}(x_i) \to \bar{\alpha}(x)$ strictly. The fact that $\bar{\alpha}$ is unital follows from the definition of $\bar{\alpha}$ (see the proof of Proposition 2.1).

(ii) $\Rightarrow$ (iii): Suppose that $\bar{\alpha}$ satisfies (ii), and let (e_i) be any approximate unit for A. It follows from the proof of Proposition 1.3 that (e_i) converges strictly to the identity in $M(A)$. So the assumed conditions on $\bar{\alpha}$ imply that $\alpha(e_i) \to 1$ strictly in $\mathcal{L}(E)$.

(iii) $\Rightarrow$ (i): If $\alpha(e_i) \to 1$ strictly, for some approximate unit (e_i) of A, then $\alpha(e_i)\xi \to \xi$ for all ξ in E, which shows that $\alpha(A)E$ is dense in E. Thus α is nondegenerate.

Note that condition (iii) above only requires $\alpha(e_i) \to 1$ strictly for some particular approximate unit (e_i). But the proof of (ii) $\Rightarrow$ (iii) shows that if (iii) holds for one approximate unit then it must also hold for any other approximate unit. Later (in Chapter 5), we shall see other instances of this phenomenon.

For α in $\mathrm{Mor}(A, B)$ we shall normally write α, rather than $\bar{\alpha}$, for the extension of α to $M(A)$. If $\beta \in \mathrm{Mor}(B, C)$ then the composition $\beta\alpha$ is defined, provided that we interpret β to be the extended mapping from $M(B)$ to $M(C)$. Obviously $\beta\alpha$ is a $*$-homomorphism from A to $M(C)$, but it is not obvious from the definition of nondegeneracy that $\beta\alpha$ is non-degenerate. However, if we apply Proposition 2.5 (with $E = B$) then it is clear that the alternative characteristic of nondegeneracy given by condition (ii) of the proposition is preserved by composition, and we conclude that $\beta\alpha \in \mathrm{Mor}(A, C)$.

Let A, B be C*-algebras and let F be a Hilbert B-module. Using Theorem 2.4 and Proposition 2.5, one sees that an element of $\mathrm{Mor}(A, \mathcal{K}(F))$ is the same thing as a nondegenerate $*$-homomorphism from A to $\mathcal{L}(F)$.

References for Chapter 2: The straightforward and self-contained approach to multipliers given here is essentially that proposed in [Wor 1]. See also [Ior], [Val], and [Wor 5]. Theorem 2.4 is from [Kas 1].

Chapter 3

Projections and unitaries

We say that a closed submodule F of a Hilbert A-module E is *complemented* if $E = F \oplus F^{\perp}$. As already emphasised in Chapter 1, a closed submodule of a Hilbert C*-module need not be complemented. If F is complemented then for each z in E we can uniquely write $z = x + y$ with x in F and y in $F^{\perp}$. Just as in the case of Hilbert spaces, the equation $x = pz$ defines a projection p in $\mathcal{L}(E)$ whose range is F. Conversely, if p is a projection in $\mathcal{L}(E)$ then the range of p is a complemented submodule of E, since it is easy to check that

$$\operatorname{ran}(p)^{\perp} = \operatorname{ran}(1-p) = \ker(p),$$

and evidently $\operatorname{ran}(p) \oplus \operatorname{ran}(1-p) = E$.

Thus by a complemented submodule of E we mean one that is *orthogonally* complemented. We say that a closed submodule F of E is *topologically complemented* if there is a closed submodule G of E with $F + G = E$, $F \cap G = \{0\}$. If F is (orthogonally) complemented then clearly F is topologically complemented; but the converse is false. For example, take $A = C([0,1])$ and let $J = \{f \in A : f(0) = 0\}$. Let $E = A \oplus J$ and let F be the submodule given by

$$F = \{(f, f) : f \in J\}.$$

Then $F^{\perp} = \{(g, -g) : g \in J\}$, from which we see that $F \oplus F^{\perp} = J \oplus J \neq E$. So F is not complemented. But the submodule $G = \{(f, 0) : f \in A\}$ of E is a topological complement for F.

If F is a topologically complemented submodule of E, with topological complement G, then for each z in E we can uniquely write $z = x+y$ with x in F and y in G. The equation $x = qz$ defines an idempotent A-linear map q on E, which is bounded (by an easy application of the closed graph theorem). Of course, q need not be selfadjoint, and indeed q need not be adjointable at all. In fact, we shall see later in this chapter (Corollary 3.3) that if $q \in \mathcal{L}(E)$ then F must necessarily be complemented. But if q is a bounded idempotent A-linear map on E whose range is complemented, it does not follow that q need be adjointable. To see this, go back to the example in the previous paragraph, and define q on $A \oplus J$ by $q(f,g) = (f-g, 0)$. Then q is bounded and idempotent, and its range is G (which is complemented since $G^{\perp} = \{(0,g): g \in J\}$). But q is not adjointable (exercise: why not?).

Our first main result in this chapter is a theorem of Miščenko [Miš] which enables one to conclude that certain submodules are complemented. We need a preparatory lemma.

LEMMA 3.1. *Suppose that E is a Hilbert C^*-module, that $t^* = t \in \mathcal{L}(E)$ and that*

$$\|tx\| \geqslant k\|x\| \qquad (x \in E) \tag{3.1}$$

for some constant $k > 0$. Then t is invertible in $\mathcal{L}(E)$.

Proof. We have to show that 0 does not belong to the spectrum $\mathrm{sp}(t)$ of t. The proof will be by contradiction, so suppose that $0 \in \mathrm{sp}(t)$. Let f be a continuous function on $\mathbf{R}$ such that

$$f(0) = 1 = \|f\|, \quad f(\lambda) = 0 \text{ whenever } |\lambda| \geqslant \tfrac{1}{2}k.$$

Using the functional calculus on the unital C*-subalgebra of $\mathcal{L}(E)$ generated by t, let $s = f(t)$. Then $\|s\| = 1$ and $\|ts\| \leqslant \frac{1}{2}k$. Choose a unit vector x in E with $\|sx\| > \frac{1}{2}$. Then $\|tsx\| \leqslant \frac{1}{2}k < k\|sx\|$, which contradicts (3.1). Hence $0 \notin \mathrm{sp}(t)$, as required.

THEOREM 3.2. *Let E, F be Hilbert A-modules and suppose that t in $\mathcal{L}(E,F)$ has closed range. Then*

(i) *$\ker(t)$ is a complemented submodule of E,*
(ii) *$\mathrm{ran}(t)$ is a complemented submodule of F,*
(iii) *the mapping $t^* \in \mathcal{L}(F,E)$ also has closed range.*

Proof. (i): Replacing F by the range of t, we may suppose that t is surjective. By the open mapping theorem, $t(E_1)$ contains a neighbourhood of 0 in F (where E_1 is the unit ball of E), and hence contains some closed ball with radius $\delta > 0$. Thus for every y in F we can find x in E with $tx = y$ and $\|x\| \leqslant \delta^{-1}\|y\|$. Then

$$\|t^*y\|^2 = \|\langle y, tt^*y\rangle\| \leqslant \|y\|\,\|tt^*y\|$$

so that

$$\|y\|^2 = \|\langle tx, y\rangle\| = \|\langle x, t^*y\rangle\| \leqslant \|x\|\,\|t^*y\| \leqslant \delta^{-1}\|y\|^{\frac{3}{2}}\|tt^*y\|^{\frac{1}{2}},$$

from which $\|y\| \leqslant \delta^{-2}\|tt^*y\|$. Thus tt^* satisfies (3.1), and it follows from Lemma 3.1 that tt^* is invertible. In particular, tt^* is surjective.

Given z in E, it follows that $tz = tt^*w$ for some w in F. Therefore $z - t^*w \in \ker(t)$; and

$$z = (z - t^*w) + t^*w \in \ker(t) + \operatorname{ran}(t^*).$$

Hence $\operatorname{ran}(t^*)$, which is easily seen to be orthogonal to $\ker(t)$, is in fact the complement of $\ker(t)$. This completes the proof of (i).

(ii) and (iii): It is tempting to argue as follows: the proof of (i) shows that $\operatorname{ran}(t^*)$ is closed. So we can apply the argument of (i) with t^* in place of t, and conclude that $\ker(t^*)$ is the (orthogonal) complement of $\operatorname{ran}(t)$.

However, there is a gap in this argument. At the start of the proof of (i), we replaced F by $F_0 = \operatorname{ran}(t)$. So the map t^* that occurs in the rest of the proof of (i) is not the adjoint of $t \in \mathcal{L}(E, F)$ but the adjoint of $t \in \mathcal{L}(E, F_0)$. To make this more precise, write $\hat{t}$ for the mapping t considered as an element of $\mathcal{L}(E, F_0)$. The proof of (i) shows that $\ker(t)$ has an orthogonal complement $\operatorname{ran}(\hat{t}^*)$. We now want to show that $\operatorname{ran}(\hat{t}^*) = \operatorname{ran}(t^*)$.

To see that this is the case, note first that, by an easy calculation, the restriction of t^* to F_0 is $\hat{t}^*$. So $\operatorname{ran}(\hat{t}^*) \subseteq \operatorname{ran}(t^*)$. On the other hand, $\operatorname{ran}(t^*) \subseteq \ker(t)^\perp = \operatorname{ran}(\hat{t}^*)$ by (i).

Putting this together with the proof of (i), we see that if $t \in \mathcal{L}(E, F)$ has closed range then t^* also has closed range, and $\ker(t)$ is the orthogonal complement of $\operatorname{ran}(t^*)$. In this statement, we really can exchange the roles of t and t^*, and conclude that $\operatorname{ran}(t)$ has an orthogonal complement $\ker(t^*)$.

COROLLARY 3.3. *If q is an idempotent in $\mathcal{L}(E)$ then the range of q is a complemented submodule of E .*

Proof. The range of a bounded idempotent mapping is always closed, so Theorem 3.2 applies.

If $t \in \mathcal{L}(E,F)$ does not have closed range then neither $\ker(t)$ nor the closure of $\operatorname{ran}(t)$ need be complemented. For example, let $A = C([-1,1])$, take $F = E = A$ and define t by

$$tf(\lambda) = \begin{cases} \lambda f(\lambda) & \text{if } 0 \leqslant \lambda \leqslant 1, \\ 0 & \text{if } -1 \leqslant \lambda < 0, \end{cases}$$

for all f in E. Then

$$\ker(t) = \{f \in E : f(\lambda) = 0 \text{ whenever } 0 \leqslant \lambda \leqslant 1\},$$
$$\overline{\operatorname{ran}(t)} = \{f \in E : f(\lambda) = 0 \text{ whenever } -1 \leqslant \lambda \leqslant 0\},$$

and neither of these submodules is complemented.

For a general t in $\mathcal{L}(E,F)$ it is easy to verify that $\operatorname{ran}(t)^{\perp} = \ker(t^*)$. But it need not be the case that $\ker(t^*)^{\perp} = \overline{\operatorname{ran}(t)}$. Example: take t to be the operation of multiplication by the independent variable on $C([0,1])$.

An operator $u \in \mathcal{L}(E,F)$ is said to be *unitary* if

$$u^*u = 1_E, \qquad uu^* = 1_F.$$

If there exists a unitary element of $\mathcal{L}(E,F)$ then we say that E and F are *unitarily equivalent* Hilbert A-modules, and we write $E \approx F$.

Exercise (suitable only for those who know enough C*-algebra theory to understand what the problem means): Show that the Hilbert A-modules A and $A \oplus A$ are unitarily equivalent if and only if $M(A)$ contains a unital C*-subalgebra isomorphic to the Cuntz algebra $\mathcal{O}_2$.

If $u \in \mathcal{L}(E,F)$ is unitary then it is clear that u is a surjective A-linear map, and also that u is isometric:

$$\|ux\| = \|x\| \qquad (x \in E). \tag{3.2}$$

Our next main result in this chapter will be the converse assertion, that if $u: E \to F$ is an isometric, surjective A-linear map then u is unitary. To understand what is involved in proving this, observe that if instead of (3.2) we had the analogous result for the A-valued "norms" on E and F, namely

$$|ux| = |x| \qquad (x \in E), \tag{3.3}$$

then we could square both sides to get $\langle ux, ux\rangle = \langle x, x\rangle$. Then by polarisation we would obtain

$$\langle ux, uy\rangle = \langle x, y\rangle \qquad (x, y \in E). \tag{3.4}$$

Since by assumption u is invertible, (3.4) shows that u is adjointable, with $u^* = u^{-1}$, which is the same as saying that u is unitary.

Thus our task is to show that (3.2) implies (3.3) for an A-linear map $u: E \to F$. As in the case of Miščenko's Theorem 3.2, we require a preliminary lemma which is an exercise in the use of the functional calculus.

LEMMA 3.4. *Suppose that a, b are positive elements of the C*-algebra A, and that $\|ac\| = \|bc\|$ for all c in A. Then $a = b$.*

Proof. We may suppose that a and b lie in the unit ball of A. Since positive elements of a C*-algebra have unique positive square roots, it will be sufficient to show that $a^2 = b^2$. We shall prove this by contradiction.

Assume then that $a^2 - b^2 \neq 0$, so that $\mathrm{sp}(a^2 - b^2)$ does not reduce to $\{0\}$. Let $\delta = \frac{1}{2}\sup \mathrm{sp}(a^2 - b^2)$. Exchanging a and b if need be, we may suppose that $\delta > 0$. Let f be a continuous real-valued function on $\mathrm{sp}(a^2 - b^2)$ satisfying

$$0 \leqslant f(\lambda) \leqslant 1 \qquad (\lambda \in \mathrm{sp}(a^2 - b^2)),$$
$$f(\lambda) = 0 \text{ whenever } \lambda \leqslant \delta, \quad f(2\delta) = 1,$$

and let $c = f(a^2 - b^2) \in A$. It follows from the functional calculus that

$$\|c(a^2 - b^2)c\| = 2\delta. \tag{3.5}$$

Since $\lambda > \delta$ whenever $f(\lambda) \neq 0$, it follows, again by use of the functional calculus, that $c(a^2 - b^2)c \geqslant \delta c^2$.

Let ρ be a state of A such that $\rho(cb^2c) = \|cb^2c\|$. Since $\|b\| \leqslant 1$, we must have $\rho(cb^2c) \leqslant \rho(c^2)$, and hence

$$\rho(ca^2c) \geqslant \rho(cb^2c + \delta c^2) \geqslant (1 + \delta)\rho(cb^2c),$$

from which $\|ca^2c\| \geqslant (1+\delta)\|cb^2c\|$. This shows that if $cb^2c \neq 0$ then $\|ca^2c\| > \|cb^2c\|$. On the other hand, if $cb^2c = 0$ then it follows from (3.5) that $ca^2c \neq 0$. So in either case we have $\|ca^2c\| > \|cb^2c\|$. Thus $\|ac\| > \|bc\|$, in contradiction to the hypothesis of the lemma.

THEOREM 3.5. *Let A be a C^*-algebra, let E, F be Hilbert A-modules and let u be a linear map from E to F. Then the following conditions are equivalent:*

(i) *u is an isometric, surjective A-linear map;*
(ii) *u is a unitary element of $\mathcal{L}(E,F)$.*

Proof. Suppose that (i) holds. For x in E and a in A, we have

$$\begin{aligned} \|\,|ux|a\| &= \|a^*\langle ux, ux\rangle a\|^{\frac{1}{2}} \\ &= \|\langle u(xa), u(xa)\rangle\|^{\frac{1}{2}} \\ &= \|u(xa)\| \\ &= \|xa\| = \|a^*\langle x, x\rangle a\|^{\frac{1}{2}} = \|\,|x|a\|. \end{aligned}$$

It follows from Lemma 3.4 that $|ux| = |x|$. Thus (3.3) holds. As already discussed, this implies that u satisfies (ii); and the implication (ii) $\Rightarrow$ (i) is obvious.

It is natural to ask whether there should be a result analogous to Theorem 3.5 for isometries. Unfortunately, if $w: E \to F$ is an isometric A-linear map then w may not be adjointable. For example, suppose that E is a submodule of F. Then the inclusion map $i: E \to F$ is an isometric A-linear map. If i is adjointable then it is easy to check that $\ker(i^*)$ is a complement for E in F. Thus a necessary condition for i to be adjointable is that its range should be complemented. The following proposition shows that this condition can be used to characterise isometries in $\mathcal{L}(E,F)$.

PROPOSITION 3.6. *With A, E, F as before, let w be a linear map from E to F. The following conditions are equivalent:*

(i) *w is an isometric A-linear map with complemented range;*
(ii) *$w \in \mathcal{L}(E,F)$ and $w^*w = 1_E$.*

Proof. Suppose that (i) holds. Just as in the proof of Theorem 3.5, we can show that

$$\langle wx, wy\rangle = \langle x, y\rangle \qquad (x,\ y \in E). \tag{3.6}$$

Since $\operatorname{ran}(w)$ is complemented, it is the range of a projection p in $\mathcal{L}(F)$. It is easy to check that $w^{-1}p$ is an adjoint for w and that (ii) holds.

Conversely, if (ii) holds then so does (3.6), from which it is clear that w is isometric; and ww^* is a projection whose kernel is a complement for $\operatorname{ran}(w)$.

We have already encountered examples of unitary equivalence between Hilbert C*-modules, and we want to draw attention to two particular examples which frequently arise, and which we will take for granted in future. The first example occurred in Chapter 1, where we stated that if A is a C*-algebra and H is a Hilbert space with orthonormal basis $\{\varepsilon_i\}$ then $H \otimes A$ "can be naturally identified with" $\bigoplus_i A_i$, where each A_i is a copy of A. What this means, of course, is that $H \otimes A \approx \bigoplus_i A_i$, the unitary that implements this equivalence being the map u that takes $\varepsilon_i \otimes a$ to the element of $\bigoplus_i A_i$ that has a in the ith coordinate and zeros elsewhere.

Our second natural example of a unitary equivalence is that if H is a Hilbert space and X is a locally compact Hausdorff space then we have $H \otimes C_0(X) \approx C_0(X, H)$ as Hilbert $C_0(X)$-modules, where $C_0(X, H)$ is the space of continuous H-valued functions on X vanishing at infinity. The $C_0(X)$-module structure and inner product on $C_0(X, H)$ are defined pointwise. (In fact, this is just the nonunital analogue of the example discussed at the outset of Chapter 1.) The unitary map $v \colon H \otimes C_0(X) \to C_0(X, H)$ is given by defining $v(\xi \otimes f)$ to be the map $\lambda \mapsto f(\lambda)\xi \quad (\lambda \in X)$, where $\xi \in H$ and $f \in C_0(X)$. It is straightforward to verify that v is unitary, and also that $\mathcal{L}(C_0(X, H))$ consists of multiplication by elements of $C_b^{\text{str}}(X, \mathcal{L}(H))$, the algebra of continuous bounded functions from X to $\mathcal{L}(H)$ with its strict (or strong*) topology. In Chapter 8, we shall need to make use of this natural correspondence between $\mathcal{L}(H \otimes C_0(X))$ and $C_b^{\text{str}}(X, \mathcal{L}(H))$. It is also straightforward to verify that under this correspondence $\mathcal{K}(H \otimes C_0(X))$ corresponds to $C_0(X, \mathcal{K}(H))$, the space of norm-continuous maps from X to $\mathcal{K}(H)$ that vanish at infinity.

It can be seen from the proof of Theorem 3.2 that if $t \in \mathcal{L}(E, F)$ has closed range then $\operatorname{ran}(t^*) = \operatorname{ran}(t^*t)$. The condition that the range of an operator should be closed is a very strong one, and we now want to examine what can be said when we drop this condition. For operators between Hilbert spaces it is always true that $\operatorname{ran}(t^*)$ and $\operatorname{ran}(t^*t)$ have the same closure (since both closures are easily seen to be the orthogonal complement of $\ker(t)$). The following proposition says that this result still holds for adjointable operators between Hilbert C*-modules (even though $\overline{\operatorname{ran}(t^*)}$ may not be complemented). Since we cannot use orthogonal complements, the proof is necessarily more elaborate than in the Hilbert space case.

PROPOSITION 3.7. *For t in $\mathcal{L}(E,F)$, t^*F and t^*tE have the same closure.*

Proof. Write J, K for the right ideals of $\mathcal{L}(E)$ given by

$$J = \overline{\{t^*tu: u \in \mathcal{L}(E)\}}, \qquad K = \overline{\{t^*v: v \in \mathcal{L}(E,F)\}}$$

(the bars denoting closure in the norm topology). Clearly JE and t^*tE have the same closure.

For x, y in E and z in F, we have $\theta_{z,x} \in \mathcal{L}(E,F)$ and $\theta_{z,x}(y) = z\langle x, y\rangle$, from which it follows that

$$F\langle E, E\rangle \subseteq \mathcal{L}(E,F)E.$$

Let (e_i) be an approximate unit for $\langle E, E\rangle^-$, and let $y \in F$. Then

$$t^*y = \lim_i (t^*y)e_i$$

(see (1.5) and the following paragraph), and

$$t^*(ye_i) \in t^*\left(F\langle E,E\rangle^-\right) \subseteq t^*\left(\mathcal{L}(E,F)E\right)^- \subseteq (KE)^-.$$

It follows that t^*F is contained in the closure of KE. Conversely, KE is evidently contained in the closure of t^*F.

So to show that t^*F and t^*tE have the same closure, it will suffice to prove that $J = K$. Clearly $J \subseteq K$, so we prove the reverse inclusion.

Let ρ be a state of $\mathcal{L}(E)$ that vanishes on J. Then in particular $\rho(t^*t) = 0$. The sesquilinear form on $\mathcal{L}(E,F)$ defined by

$$(u, v) \mapsto \rho(u^*v)$$

satisfies the Cauchy–Schwarz inequality. Therefore

$$|\rho(t^*u)|^2 \leqslant \rho(t^*t)\rho(u^*u) = 0 \qquad (u \in \mathcal{L}(E, F)),$$

and so ρ vanishes on K. Thus every state of $\mathcal{L}(E)$ that vanishes on J also vanishes on K. By 2.9.4 of [Dix 2] or 10.2.9 of [KadRin], it follows that $J = K$.

Adjointable operators between Hilbert C*-modules do not generally have a polar decomposition. This is because partial isometries between Hilbert C*-modules are not adjointable unless their initial and final spaces are complemented (as we have already seen, in the case of isometries, in the discussion before Proposition 3.6). But if t and t^* both have dense range then this difficulty can be circumvented, and this is the idea in the proof of the following proposition.

PROPOSITION 3.8. *If E, F are Hilbert A-modules and there is an element t of $\mathcal{L}(E, F)$ such that t and t^* have dense range then $E \approx F$.*

Proof. Since t^*t has dense range (by Proposition 3.7), so does the operator $|t| = (t^*t)^{\frac{1}{2}}$. Define $u\colon \operatorname{ran}(t) \to \operatorname{ran}(|t|)$ by

$$u(tx) = |t|x \qquad (x \in E).$$

For x, y in E, we have

$$\langle utx, uty \rangle = \langle x, |t|^2 y \rangle = \langle tx, ty \rangle.$$

So u is isometric and, since t and $|t|$ both have dense range, u extends to a unitary map from F to E.

We conclude this chapter with a brief indication of how the idea in Proposition 3.8 can be used to construct a polar decomposition for certain adjointable operators.

An element c in $\mathcal{L}(E, F)$ is called a *partial isometry* (from E_0 to F_0) if $F_0 = \operatorname{ran}(c)$ is complemented in F and there exists a complemented submodule E_0 of E such that c is isometric from E_0 onto F_0 and $c(E_0^\perp) = \{0\}$. Just as for Hilbert space operators ([KadRin], Proposition 6.1.1; [Mur],

Theorem 2.3.3), one can easily check that the following conditions are equivalent, for an element c of $\mathcal{L}(E,F)$:

(i) c is a partial isometry,
(ii) c^*c is a projection in $\mathcal{L}(E)$,
(iii) cc^* is a projection in $\mathcal{L}(F)$,
(iv) $cc^*c = c$,
(v) $c^*cc^* = c^*$.

Suppose that $t \in \mathcal{L}(E,F)$ and that the closures of the ranges of t and t^* are both complemented. Then the proof of Proposition 3.8 shows that t has a polar decomposition $t = c|t|$, where $c \in \mathcal{L}(E,F)$ is a partial isometry.

References for Chapter 3: [Miš], [Lan 1], [Wor 5].

Chapter 4

Tensor products

We shall assume some familiarity with the theory of tensor products of Hilbert spaces and C*-algebras, as presented in [KadRin] or [Mur] for example. If H, K are Hilbert spaces and A, B are C*-algebras then we write $H \otimes_{\text{alg}} K$ and $A \otimes_{\text{alg}} B$ for their algebraic (linear space) tensor products. We denote by $H \otimes K$ the Hilbert space tensor product of H and K, namely the completion of $H \otimes_{\text{alg}} K$ with respect to the inner product given on simple tensors by

$$\langle \xi_1 \otimes \eta_1, \xi_2 \otimes \eta_2 \rangle = \langle \xi_1, \xi_2 \rangle \langle \eta_1, \eta_2 \rangle \qquad (\xi_1, \xi_2 \in H,\ \eta_1, \eta_2 \in K).$$

For C*-algebras, we denote by $A \otimes B$ the completion of $A \otimes_{\text{alg}} B$ with respect to the spatial, or minimal, C*-norm. This is the only C*-algebraic tensor product that we shall consider, and we shall need the following properties, which are discussed in the books referred to above.

(i) The algebraic tensor product $A^* \otimes_{\text{alg}} B^*$ of the (Banach) dual spaces of A and B embeds in $(A \otimes B)^*$. In other words, if $f \in A^*$ and $g \in B^*$ then the linear functional $f \otimes g$ on $A \otimes_{\text{alg}} B$ is continuous for the C*-norm and therefore extends to a (bounded) linear functional on $A \otimes B$.

(ii) For r in $A \otimes_{\text{alg}} B$, the norm of r is given by the following formula, in which $S(A)$, $S(B)$ denote the state spaces of A, B:

$$\|r\|^2 = \sup\left\{ \frac{(\rho \otimes \sigma)(t^* r^* r t)}{(\rho \otimes \sigma)(t^* t)} : \rho \in S(A),\ \sigma \in S(B),\ t \in A \otimes_{\text{alg}} B,\ (\rho \otimes \sigma)(t^* t) \neq 0 \right\}. \tag{4.1}$$

(iii) If π, π' are faithful representations of A, B on Hilbert spaces H, K respectively, then $\pi \otimes \pi'$ is a faithful representation of $A \otimes B$ on $H \otimes K$.

(iv) The GNS representation of $\rho \otimes \sigma$ is unitarily equivalent to $\pi_\rho \otimes \pi'_\sigma$, where π_ρ is the GNS representation of $\rho \in S(A)$ and π'_σ is the GNS representation of $\sigma \in S(B)$.

In this chapter we shall study tensor products of Hilbert C*-modules. In fact, there are two useful ways in which one can form tensor products of Hilbert C*-modules. The first, usually called the exterior tensor product, is the construction that one would expect from the algebraic theory of tensor products. The second construction, known as the interior tensor product, looks less natural but is actually more useful in applications. Before describing either of these constructions, however, we need to prove some lemmas that are mainly concerned with order properties of matrices of adjointable operators.

The first of the following lemmas is in [Pas]. The ideas in Lemmas 4.2 and 4.3 have been around for a long time (see for example Exercise 11.5.20 in [KadRin]).

LEMMA 4.1. *Let E be a Hilbert A-module and let t be a bounded A-linear operator on E. The following conditions are equivalent:*

(i) *t is a positive element of $\mathcal{L}(E)$;*

(ii) *$\langle x, tx \rangle \geqslant 0$ for all x in E.*

Proof. If $t \geqslant 0$ in $\mathcal{L}(E)$ then $\langle x, tx \rangle = |t^{\frac{1}{2}}x|^2 \geqslant 0$ in A. Conversely, if $\langle x, tx \rangle \geqslant 0$ (all x in E) then it follows by polarisation that t is selfadjoint (in particular, t is adjointable!), so we can write $t = r - s$ where r, s are positive elements of $\mathcal{L}(E)$ with $rs = 0$. For x in E, we have $\langle sx, tsx \rangle \geqslant 0$, from which $-\langle x, s^3x \rangle \geqslant 0$. But $s^3 \geqslant 0$, so that $\langle x, s^3x \rangle = 0$. Therefore s^3, and hence s, is zero, and $t = r \geqslant 0$.

If A is a C*-algebra and n is a positive integer then we write $M_n(A)$ for the C*-algebra of $n \times n$ matrices over A; and we write M_n for $M_n(\mathbf{C})$. Recall from Chapter 1 that if E is a Hilbert A-module then $\mathcal{L}(E^n)$ is naturally identified with $M_n(\mathcal{L}(E))$, and $\mathcal{K}(E^n)$ with $M_n(\mathcal{K}(E))$.

LEMMA 4.2. *Let E be a Hilbert A-module. If $x_1, \ldots, x_n \in E$ then $X \geqslant 0$*

in $M_n(A)$, where X is the matrix with (i,j)-entry $\langle x_i, x_j\rangle$. Also, if $t \in \mathcal{L}(E)$ and W is the matrix with (i,j)-entry $\langle tx_i, tx_j\rangle$ then $W \leqslant \|t\|^2 X$.

Proof. Identifying $M_n(A)$ with $\mathcal{K}(A^n)$, we have, for all $a = (a_1, \ldots, a_n)$ in A^n,

$$\langle a, Xa\rangle = \sum_{i,j} a_i^*\langle x_i, x_j\rangle a_j = \Big|\sum_i x_i a_i\Big|^2 \geqslant 0,$$

and so $X \geqslant 0$ by Lemma 4.1.

Also,

$$\begin{aligned}\langle a, Wa\rangle &= \sum_{i,j} a_i^*\langle tx_i, tx_j\rangle a_j \\ &= \Big|\sum_i tx_i a_i\Big|^2 \\ &\leqslant \|t\|^2\Big|\sum_i x_i a_i\Big|^2 = \|t\|^2\langle a, Xa\rangle,\end{aligned}$$

the inequality coming from Proposition 1.2. Thus

$$\langle a, (\|t\|^2 X - W)a\rangle \geqslant 0 \qquad (a \in A^n),$$

and so $W \leqslant \|t\|^2 X$ by Lemma 4.1 again.

LEMMA 4.3. *Let A, B be C^*-algebras and n a positive integer. Suppose that $a = (a_{ij}) \geqslant 0$ in $M_n(A)$ and $b = (b_{ij}) \geqslant 0$ in $M_n(B)$. Then*

(i) *$(a_{ij} \otimes b_{ij}) \geqslant 0$ in $M_n(A \otimes B)$;*

(ii) *$\sum_{i,j} a_{ij} \otimes b_{ij} \geqslant 0$ in $A \otimes B$;*

(iii) *if $a \leqslant c = (c_{ij}) \in M_n(A)$ and $b \leqslant d = (d_{ij}) \in M_n(B)$ then*

$$(a_{ij} \otimes b_{ij}) \leqslant (c_{ij} \otimes d_{ij})$$

in $M_n(A \otimes B)$.

Proof. (i): Since $a \geqslant 0$, we have $a = s^*s$ for some $s = (s_{ij})$ in $M_n(A)$. Thus $a_{ij} = \sum_k s_{ki}^* s_{kj}$. Similarly $b_{ij} = \sum_k t_{ki}^* t_{kj}$ in B. So (writing a_i for s_{ki} and b_i for t_{ki}) it suffices to prove that $(a_i^* a_j \otimes b_i^* b_j) \geqslant 0$ in $M_n(A \otimes B)$, for all $a_1, \ldots, a_n$ in A and $b_1, \ldots, b_n$ in B. But $a_i^* a_j \otimes b_i^* b_j = (a_i \otimes b_i)^*(a_j \otimes b_j)$, so the result follows by applying Lemma 4.2 to the Hilbert $(A \otimes B)$-module $A \otimes B$, with $x_i = a_i \otimes b_i$.

(ii): Let (u_α), (v_β) be approximate units for A, B respectively and let

$$e_{\alpha\beta} = (u_\alpha \otimes v_\beta, \dots, u_\alpha \otimes v_\beta) \in (A \otimes B)^n.$$

It follows from (i) that $\langle e_{\alpha\beta}, (a_{ij} \otimes b_{ij}) e_{\alpha\beta} \rangle \geqslant 0$, and the result follows on taking limits over α, β.

(iii): By (i), the matrices $\big((c_{ij} - a_{ij}) \otimes b_{ij}\big)$ and $\big(c_{ij} \otimes (d_{ij} - b_{ij})\big)$ are both positive in $M_n(A \otimes B)$. Hence so is their sum.

We are now in a position to describe the construction of the exterior tensor product of Hilbert C*-modules. Suppose that A, B are C*-algebras, E is a Hilbert A-module and F is a Hilbert B-module. We want to define $E \otimes F$ as a Hilbert $(A \otimes B)$-module. Start by forming the algebraic tensor product $E \otimes_{\text{alg}} F$ of the vector spaces E and F (over $\mathbf{C}$). This is a right module over $A \otimes_{\text{alg}} B$ (the module action being given by $(x \otimes y)(a \otimes b) = xa \otimes yb$). For x_1, x_2 in E and y_1, y_2 in F, we define

$$\langle x_1 \otimes y_1, x_2 \otimes y_2 \rangle = \langle x_1, x_2 \rangle \otimes \langle y_1, y_2 \rangle.$$

This extends by linearity to an $(A \otimes_{\text{alg}} B)$-valued sesquilinear form on $E \otimes_{\text{alg}} F$, which makes $E \otimes_{\text{alg}} F$ into a semi-inner-product module over the pre-C*-algebra $A \otimes_{\text{alg}} B$: to see that it is positive semi-definite, note that if $z = \sum_i x_i \otimes y_i \in E \otimes_{\text{alg}} F$ then

$$\langle z, z \rangle = \sum_{i,j} \langle x_i, x_j \rangle \otimes \langle y_i, y_j \rangle.$$

By Lemma 4.2, the matrix $\big(\langle x_i, x_j \rangle\big)$ is positive in $M_n(A)$, and $\big(\langle y_i, y_j \rangle\big) \geqslant 0$ in $M_n(B)$. So it follows from Lemma 4.3(ii) that $\langle z, z \rangle \geqslant 0$, as required.

The semi-inner product on $E \otimes_{\text{alg}} F$ is actually an inner product. To prove that it is positive definite is quite hard: for if $z \in E \otimes_{\text{alg}} F$ and $\langle z, z \rangle = 0$ in $A \otimes_{\text{alg}} B$, then to conclude that $z = 0$ in $E \otimes_{\text{alg}} F$ we need to pull back information from $A \otimes_{\text{alg}} B$ to $E \otimes_{\text{alg}} F$, and there is no easy way to do this. We therefore defer the proof of this until Chapter 6, leaving it to the conscientious reader to check in due course that this forward reference does not lead to any circular arguments. Accepting for the time being that this fact has been established, we see that $E \otimes_{\text{alg}} F$ is an inner-product module over the pre-C*-algebra $A \otimes_{\text{alg}} B$, and we can perform the double completion process discussed in Chapter 1 to conclude that the completion $E\otimes F$ of $E\otimes_{\text{alg}}F$ is a Hilbert $(A\otimes B)$-module. We call $E\otimes F$ the *exterior tensor product* of E and F.

If H is a Hilbert space (that is, a Hilbert $\mathbb{C}$-module) and A is a C*-algebra regarded as a Hilbert module over itself, then the above construction defines $H\otimes A$ as a Hilbert $(\mathbb{C}\otimes A)$-module. Of course, $\mathbb{C}\otimes A$ is naturally isomorphic to A, and we leave it to the reader to check that the Hilbert A-module obtained in this way is the same as (or at least unitarily equivalent to) the Hilbert A-module $H\otimes A$ as defined in Chapter 1.

With E, F as above, we wish to investigate the adjointable operators on $E\otimes F$. Suppose that $s\in\mathcal{L}(E)$, $t\in\mathcal{L}(F)$. Define a linear operator $s\otimes t$ on $E\otimes_{\text{alg}}F$ by

$$s\otimes t\cdot x\otimes y = sx\otimes ty \qquad (x\in E,\ y\in F).$$

If $z=\sum_i x_i\otimes y_i\in E\otimes_{\text{alg}}F$ then

$$\begin{aligned}|(s\otimes t)z|^2 &= \sum_{i,j}\langle sx_i\otimes ty_i, sx_j\otimes ty_j\rangle\\ &= \sum_{i,j}\langle sx_i, sx_j\rangle\otimes\langle ty_i, ty_j\rangle\\ &\leqslant \|s\|^2\|t\|^2\sum_{i,j}\langle x_i, x_j\rangle\otimes\langle y_i, y_j\rangle\end{aligned}$$

by Lemmas 4.2 and 4.3. (Indeed, $(\langle sx_i, sx_j\rangle)\leqslant\|s\|^2(\langle x_i,x_j\rangle)$ in $M_n(A)$, and a similar inequality with t and (y_i) holds in $M_n(B)$. Therefore

$$\|s\|^2\|t\|^2(\langle x_i,x_j\rangle\otimes\langle y_i,y_j\rangle)-(\langle sx_i,sx_j\rangle\otimes\langle ty_i,ty_j\rangle)$$

is a positive element of $M_n(A\otimes B)$ by Lemma 4.3(iii). Now argue as in the proof of Lemma 4.3(ii) to show that the sum of the elements of this matrix is positive in $A\otimes B$.)

Thus $|(s\otimes t)z|^2\leqslant\|s\|^2\|t\|^2|z|^2$ and so $\|(s\otimes t)z\|\leqslant\|s\|\,\|t\|\,\|z\|$. Hence $s\otimes t$ is bounded on $E\otimes_{\text{alg}}F$ and extends by continuity to a bounded linear operator (which we still call $s\otimes t$) on $E\otimes F$. It is a routine verification that $s^*\otimes t^*$ is an adjoint for $s\otimes t$, so in fact $s\otimes t\in\mathcal{L}(E\otimes F)$. The map $(s,t)\mapsto s\otimes t$ is clearly bilinear, and therefore defines a linear map j from $\mathcal{L}(E)\otimes_{\text{alg}}\mathcal{L}(F)$ into $\mathcal{L}(E\otimes F)$. It is straightforward to check that j is a $*$-homomorphism.

We show next that j is continuous for the C*-norm on $\mathcal{L}(E)\otimes\mathcal{L}(F)$ and so extends to a $*$-homomorphism from $\mathcal{L}(E)\otimes\mathcal{L}(F)$ into $\mathcal{L}(E\otimes F)$. To see this, let $r = \sum_i s_i \otimes t_i$ be a positive element of $\mathcal{L}(E)\otimes_{\text{alg}}\mathcal{L}(F)$. Given $\varepsilon > 0$, we can choose a unit vector $z = \sum_p x_p \otimes y_p$ in $E\otimes_{\text{alg}}F$ with

$$\|\langle z, j(r)z\rangle\| > \|j(r)\| - \varepsilon.$$

But $\langle z, j(r)z\rangle$ is a positive element of $A\otimes_{\text{alg}}B$, so by (4.1) there exist states ρ of A and σ of B, and an element $d = \sum_k a_k \otimes b_k$ of $A\otimes_{\text{alg}}B$, such that

$$\frac{(\rho\otimes\sigma)(d^*\langle z, j(r)z\rangle d)}{(\rho\otimes\sigma)(d^*d)} > \|j(r)\| - \varepsilon. \tag{4.2}$$

But

$$\begin{aligned}(\rho\otimes\sigma)(d^*\langle z, j(r)z\rangle d) &= \sum_{p,k,i,q,l} \rho\big(\langle x_p a_k, s_i x_q a_l\rangle\big)\, \sigma\big(\langle y_p b_k, t_i y_q b_l\rangle\big) \\ &= \sum_\alpha (f_\alpha \otimes g_\alpha)(r),\end{aligned}$$

where $\alpha = (p, k, q, l)$,

$$\begin{aligned} f_\alpha(s) &= \rho\big(\langle x_p a_k, s x_q a_l\rangle\big) \qquad (s \in \mathcal{L}(E)), \\ g_\alpha(t) &= \sigma\big(\langle y_p b_k, t y_q b_l\rangle\big) \qquad (t \in \mathcal{L}(F)). \end{aligned}$$

Since $\sum_\alpha f_\alpha \otimes g_\alpha$ is in $\mathcal{L}(E)^*\otimes_{\text{alg}}\mathcal{L}(F)^*$, this functional is continuous for the C*-norm on $\mathcal{L}(E)\otimes\mathcal{L}(F)$; and since it is a positive linear functional on $\mathcal{L}(E)\otimes_{\text{alg}}\mathcal{L}(F)$, we must have the inequality

$$\sum_\alpha (f_\alpha \otimes g_\alpha)(r) \leqslant \|r\| \sum_\alpha (f_\alpha \otimes g_\alpha)(1)$$

for the norm of r in the C*-tensor product $\mathcal{L}(E)\otimes\mathcal{L}(F)$. In other words,

$$(\rho\otimes\sigma)(d^*\langle z, j(r)z\rangle d) \leqslant \|r\|(\rho\otimes\sigma)(d^*\langle z, z\rangle d) \leqslant \|r\|(\rho\otimes\sigma)(d^*d)$$

(the last inequality coming from the fact that $\|z\| = 1$). It follows from (4.2) that $\|j(r)\| - \varepsilon < \|r\|$. Since ε was arbitrary we conclude that j is norm-reducing on positive elements of $\mathcal{L}(E)\otimes_{\text{alg}}\mathcal{L}(F)$ and hence on all elements (by the C*-condition). Thus j extends by continuity to a $*$-homomorphism from $\mathcal{L}(E)\otimes\mathcal{L}(F)$ into $\mathcal{L}(E\otimes F)$.

In fact, j is injective and is therefore an isometric embedding. This is proved as follows. If ρ is a state of A, and x is an element of E with $\rho(\langle x,x\rangle)=1$, then the map $s\mapsto\rho(\langle x,sx\rangle)$ gives a state of $\mathcal{L}(E)$. For s in $\mathcal{L}(E)$ we have

$$\|s\|^2=\sup_{\rho,x}\rho(\langle x,s^*sx\rangle),$$

so if $\pi_{\rho,x}$ is the GNS representation associated with the above state of $\mathcal{L}(E)$ then $\bigoplus_{\rho,x}\pi_{\rho,x}=\pi_E$ is a faithful $*$-representation of $\mathcal{L}(E)$. Similarly, for each state σ of B and each y in F with $\sigma(\langle y,y\rangle)=1$ we can form a representation $\pi'_{\sigma,y}$ of $\mathcal{L}(F)$, and $\pi'_F=\bigoplus_{\sigma,y}\pi'_{\sigma,y}$ is faithful. Therefore

$$\pi_E\otimes\pi'_F\approx\bigoplus_{\rho,\sigma,x,y}\pi_{\rho,x}\otimes\pi'_{\sigma,y}$$

is a faithful representation of $\mathcal{L}(E)\otimes\mathcal{L}(F)$. But $\pi_{\rho,x}\otimes\pi'_{\sigma,y}$ is (unitarily equivalent to) the GNS representation of $\mathcal{L}(E)\otimes\mathcal{L}(F)$ associated with the state

$$r\mapsto(\rho\otimes\sigma)(\langle x\otimes y,j(r)\cdot x\otimes y\rangle).$$

So if $j(r)=0$ it follows that $(\pi_E\otimes\pi'_F)(r)=0$ and thus $r=0$.

The embedding $j:\mathcal{L}(E)\otimes\mathcal{L}(F)\to\mathcal{L}(E\otimes F)$ will not in general be surjective. Even in the case when $A=B=\mathbb{C}$, so that E and F are just Hilbert spaces, $j(\mathcal{L}(E)\otimes\mathcal{L}(F))$ is only equal to $\mathcal{L}(E\otimes F)$ if at least one of E, F is finite-dimensional ([KadRin], Exercise 11.5.7). However, it is straightforward to see that $j(\mathcal{K}(E)\otimes\mathcal{K}(F))=\mathcal{K}(E\otimes F)$. This follows from the easily verified fact that if $u,v\in E$ and $x,y\in F$ then

$$j(\theta_{u,v}\otimes\theta_{x,y})=\theta_{u\otimes x,v\otimes y}.$$

In future we shall ordinarily suppress the notation j for the embedding of $\mathcal{L}(E)\otimes\mathcal{L}(F)$ in $\mathcal{L}(E\otimes F)$, and simply regard $\mathcal{L}(E)\otimes\mathcal{L}(F)$ as a C*-subalgebra of $\mathcal{L}(E\otimes F)$. In particular, in the case where $E=A$ and $F=B$, we have a canonical embedding of $M(A)\otimes M(B)$ in $M(A\otimes B)$. This embedding will not in general be surjective. (Exercise: give a counterexample.)

Suppose that A, A', B, B' are C*-algebras and that $\alpha\in\operatorname{Mor}(A,A')$, $\beta\in\operatorname{Mor}(B,B')$. Then we can form the $*$-homomorphism

$$\alpha\otimes\beta:A\otimes B\to M(A')\otimes M(B')$$

([KadRin], Theorem 11.1.3), which we regard as a map into $M(A'\otimes B')$. To show that $\alpha\otimes\beta \in \mathrm{Mor}(A\otimes B, A'\otimes B')$ we must prove that it is a nondegenerate mapping. Let (u_i), (v_j) be approximate units for A, B respectively. Then $(u_i\otimes v_j)$ (with the product ordering) is an approximate unit for $A\otimes B$. By Proposition 2.5, $\alpha(u_i) \to 1$ strictly in $M(A')$ and $\beta(v_j) \to 1$ strictly in $M(B')$. Therefore

$$\alpha(u_i)\otimes\beta(v_j)\cdot x \xrightarrow{i,j} x$$

for all x in $A'\otimes_{\mathrm{alg}}B'$ and hence by continuity for all x in $A'\otimes B'$. That is, $(\alpha\otimes\beta)(u_i\otimes v_j) \to 1$ strictly in $M(A'\otimes B')$. By Proposition 2.5 again, this shows that $\alpha\otimes\beta$ is nondegenerate.

That concludes our treatment of the exterior tensor product. The idea of the interior tensor product is this: suppose as before that E is a Hilbert A-module and F is a Hilbert B-module, and suppose in addition that we are given a $*$-homomorphism $\phi: A \to \mathcal{L}(F)$. Then we shall construct a space $E\otimes_\phi F$ having the structure of a Hilbert B-module.

In preparation for the construction that follows, we first prove a factorisation result for elements of a Hilbert C*-module.

LEMMA 4.4. *Suppose that E is a Hilbert A-module, $x \in E$ and $0 < \alpha < 1$. Then there is an element w of E such that $x = w|x|^\alpha$.*

Proof. For any continuous function f on the spectrum of $|x|$, we have

$$\begin{aligned}\big\|xf(|x|)\big\| &= \big\|f(|x|)^*\langle x,x\rangle f(|x|)\big\|^{\frac{1}{2}}\\ &= \big\||x|f(|x|)\big\|\\ &= \sup\big\{|\lambda f(\lambda)|: \lambda \in \mathrm{sp}(|x|)\big\}.\end{aligned}$$

For $n \geqslant 1$, define a function g_n by

$$g_n(\lambda) = \begin{cases} n^\alpha & \text{if } \lambda \leqslant 1/n,\\ \lambda^{-\alpha} & \text{if } \lambda > 1/n.\end{cases}$$

Using the above norm estimate, it is easy to check that $\big(xg_n(|x|)\big)$ is a Cauchy sequence in E and so has a limit w. Adjoining an identity to A if necessary and using the fact that $x1 = x$ (see Chapter 1), we have

$$\big\|xg_n(|x|)|x|^\alpha - x\big\| = \big\|x\big(g_n(|x|)|x|^\alpha - 1\big)\big\|$$

$$= \sup\{|\lambda(g_n(\lambda)\lambda^\alpha - 1)| : \lambda \in \mathrm{sp}(|x|)\}$$
$$\to 0 \text{ as } n \to \infty.$$

Therefore $w|x|^\alpha = x$, as required.

It will be convenient in the following construction, and subsequently, to use the notion of complete positivity. If $\rho: A \to B$ is a linear mapping between C*-algebras and n is a positive integer, we write $\rho^{(n)}$ for the map from $M_n(A)$ to $M_n(B)$ obtained by applying ρ to each matrix element: $\rho^{(n)}((a_{ij})) = (\rho(a_{ij}))$. We say that ρ is *completely positive* (see [KadRin], Exercise 11.5.16 or [Pau], p.5) if all the maps $\rho^{(n)}$ $(n \geqslant 1)$ are positive. Every $*$-homomorphism between C*-algebras is completely positive. In Chapter 5, we shall meet another important class of mappings (retractions) that are completely positive. For further information about completely positive mappings, see [Pau].

Let E be a Hilbert A-module. In Chapter 1, we defined E^n as a Hilbert A-module. However, E^n can also be regarded as an inner-product $M_n(A)$-module, in the following way. If $x = (x_1, \ldots, x_n)$ and $y = (y_1, \ldots, y_n)$ are elements of E^n and $a = (a_{ij}) \in M_n(A)$ then we define xa to be the "matrix product" of x and a, so that $xa = y$ means that $y_j = \sum_i x_i a_{ij}$. (In other words, we are now regarding x as a row vector rather than, as in Chapter 1, a column vector.) The inner product is given by $\langle x, y\rangle = (\langle x_i, y_j\rangle)$. Verification of the axioms (1.1) is routine, except that one needs Lemma 4.2 to show that $\langle x, x\rangle \geqslant 0$. We shall make use of this construction in the proof of Proposition 4.5 below.

Note that the two norms on E^n, given by the A-valued and the $M_n(A)$-valued inner products, are different. But these norms are equivalent (exercise: prove this, using the equivalence of the l^1 and l^∞ norms on $\mathbb{C}^n$), and so in particular E^n is complete for the $M_n(A)$-valued inner product and is therefore a Hilbert $M_n(A)$-module. The spaces of adjointable operators $\mathcal{L}_A(E^n)$ and $\mathcal{L}_{M_n(A)}(E^n)$ are isomorphic (and therefore isometric as C*-algebras), since each of them can be identified with $M_n(\mathcal{L}(E))$. We shall return to this phenomenon at the end of Chapter 5.

Now we are ready to construct the interior tensor product. Suppose that E is a Hilbert A-module, F is a Hilbert B-module and $\phi: A \to \mathcal{L}(F)$ is a $*$-homomorphism. We can regard F as a left A-module, the action being given by

$$(a, y) \mapsto \phi(a)y \qquad (a \in A,\ y \in F),$$

and we can form the algebraic tensor product of E and F over A, $E \otimes_A F$, which is a right B-module (see Section 9.5 of [Pie]). It is the quotient of the vector space tensor product $E \otimes_{\text{alg}} F$ by the subspace generated by elements of the form

$$xa \otimes y - x \otimes \phi(a)y \qquad (x \in E,\ y \in F,\ a \in A), \tag{4.3}$$

the action of B being given by

$$(x \otimes y)b = x \otimes (yb) \qquad (x \in E,\ y \in F,\ b \in B).$$

PROPOSITION 4.5. *With A, B, E, F and ϕ as above, $E \otimes_A F$ is an inner-product B-module under the inner product given on simple tensors by*

$$\langle x_1 \otimes y_1, x_2 \otimes y_2 \rangle = \langle y_1, \phi(\langle x_1, x_2 \rangle) y_2 \rangle \qquad (x_1, x_2 \in E,\ y_1, y_2 \in F). \tag{4.4}$$

Proof. The above formula extends by linearity to give a B-valued sesquilinear form on $E \otimes_{\text{alg}} F$. If $z = \sum_{i=1}^n x_i \otimes y_i \in E \otimes_{\text{alg}} F$ then

$$\begin{aligned} \langle z, z \rangle &= \sum_{i,j} \langle y_i, \phi(\langle x_i, x_j \rangle) y_j \rangle \\ &= \langle y, \phi^{(n)}(X) y \rangle, \end{aligned}$$

where $y = (y_1, \ldots, y_n) \in F^n$ and X is the element of $M_n(A)$ with matrix entries $\langle x_i, x_j \rangle$. By Lemma 4.2, $X \geqslant 0$, and since ϕ is completely positive it follows that $\langle z, z \rangle \geqslant 0$.

Thus we have shown that the sesquilinear form given by (4.4) makes $E \otimes_{\text{alg}} F$ into a semi-inner-product B-module. Let

$$N = \{z \in E \otimes_{\text{alg}} F : \langle z, z \rangle = 0\}.$$

We want to identify N with the subspace of $E \otimes_{\text{alg}} F$ generated by elements of the form (4.3). One way round, this is easy: if z is of the form (4.3) then a straightforward calculation shows that $\langle z, z \rangle = 0$. To prove the converse implication, let $z = \sum_i x_i \otimes y_i \in N$. Using the notation of the previous paragraph, we have

$$\langle y, \phi^{(n)}(X) y \rangle = 0.$$

Let $T = \phi^{(n)}(X)$. Then $T \geqslant 0$ in $M_n(\mathcal{L}(F)) \cong \mathcal{L}(F^n)$, and $T^{\frac{1}{2}}y = 0$. Since $|T^{\frac{1}{4}}y|^2 = \langle y, T^{\frac{1}{2}}y\rangle$, it follows that $T^{\frac{1}{4}}y = 0$.

In the Hilbert $M_n(A)$-module E^n let $x = (x_1, \ldots, x_n)$. Then $|x| = X^{\frac{1}{2}}$. By Lemma 4.4 there is an element $w = (w_1, \ldots, w_n)$ of E^n such that $wX^{\frac{1}{4}} = x$. If we write c_{ij} for the matrix elements of $X^{\frac{1}{4}}$ then we have $T^{\frac{1}{4}} = \phi^{(n)}(X^{\frac{1}{4}})$, so $T^{\frac{1}{4}}$ has matrix elements $(\phi(c_{ij}))$. Therefore

$$x_j = \sum_i w_i c_{ij}, \qquad \sum_j \phi(c_{ij}) y_j = 0,$$

from which it follows that

$$z = \sum_{i,j} \big(w_i c_{ij} \otimes y_j - w_i \otimes \phi(c_{ij}) y_j\big).$$

This exhibits z as a sum of elements of the form (4.3), as required.

Since by definition $E \otimes_A F = (E \otimes_{\text{alg}} F)/N$, we have now shown that (4.4) defines a B-valued inner product on $E \otimes_A F$, and this completes the proof of the proposition.

The completion of the inner-product B-module $E \otimes_A F$ is a Hilbert B-module which is denoted by $E \otimes_\phi F$ and is called the *interior tensor product* of E and F (using ϕ). We have already come across one example of an interior tensor product. This is when $F = B$ and $A = \mathbb{C}$, so that E is a Hilbert space, and we write H in place of E. Let $\iota\colon \mathbb{C} \to M(B)$ be the natural embedding of $\mathbb{C}$ as multiples of the identity in $M(B)$. Then $H \otimes_\iota B$ is (unitarily equivalent to) the Hilbert B-module $H \otimes B$.

The interior tensor product $E \otimes_\phi F$ is by definition the completion of $E \otimes_A F$, the elements of which are equivalence classes $z + N$, where $z \in E \otimes_{\text{alg}} F$ and N is as in the proof of Proposition 4.5. It will sometimes be desirable to use the notation $x \dot{\otimes} y$ for the element $x \otimes y + N$ of $E \otimes_\phi F$ (where $x \in E$, $y \in F$), and on other occasions it will be more convenient just to write this element as $x \otimes y$. Thus if we were being punctilious (ignore the pun) then we would write the left-hand side of (4.4) as $\langle x_1 \dot{\otimes} y_1, x_2 \dot{\otimes} y_2\rangle$.

An important special case of the interior tensor product occurs when $E = H \otimes A$, where H is a (separable, infinite-dimensional) Hilbert space. Using a notation that was introduced in Chapter 1, we write $H_A = H \otimes A$. Let F be a Hilbert B-module and suppose that $\phi \in \operatorname{Mor}(A, \mathcal{K}(F))$. In other words, not only is $\phi: A \to \mathcal{L}(F)$ a $*$-homomorphism (as before) but in addition ϕ satisfies the nondegeneracy condition that $\phi(A)F$ is dense in F. Since sums of elements of the form $(\xi \otimes a)\dot{\otimes} y$ (where $\xi \in H$, $a \in A$, $y \in F$) are dense in $H_A \otimes_\phi F$, and sums of elements of the form $\xi \otimes (\phi(a)y)$ are dense in $H \otimes F$, the calculation

$$\begin{aligned} \Big\| \sum_i (\xi_i \otimes a_i)\dot{\otimes} y_i \Big\|^2 &= \Big\| \sum_{i,j} \langle \xi_i, \xi_j \rangle \langle y_i, \phi(a_i^* a_j) y_j \rangle \Big\| \\ &= \Big\| \sum_i \xi_i \otimes \phi(a_i) y_i \Big\|^2 \end{aligned} \tag{4.5}$$

shows that the map $(\xi \otimes a)\dot{\otimes} y \mapsto \xi \otimes \phi(a) y$ extends linearly to a unitary mapping, and therefore $H_A \otimes_\phi F \approx H \otimes F$. In particular (when $F = B$), if $\phi \in \mathrm{Mor}(A, B)$ then $H_A \otimes_\phi B \approx H_B$. A similar but simpler calculation (with $\mathbb{C}$ in place of H) shows that if $\phi \in \mathrm{Mor}(A, \mathcal{K}(F))$ then $A \otimes_\phi F \approx F$, and in particular if $\phi \in \mathrm{Mor}(A, B)$ then $A \otimes_\phi B \approx B$.

Suppose that E is a Hilbert A-module, $t \in \mathcal{L}(E)$, F is a Hilbert B-module and $\phi: A \to \mathcal{L}(F)$ is a $*$-homomorphism. The map defined on simple tensors by $x \otimes y \mapsto tx \otimes y$ extends to a linear map $t \otimes \iota$ on $E \otimes_{\mathrm{alg}} F$. We have

$$\begin{aligned} \Big\| \sum_i t x_i \dot{\otimes} y_i \Big\|^2 &= \Big\| \sum_{i,j} \langle y_i, \phi(\langle t x_i, t x_j \rangle) y_j \rangle \Big\| \\ &\leqslant \|t\|^2 \Big\| \sum_{i,j} \langle y_i, \phi(\langle x_i, x_j \rangle) y_j \rangle \Big\| \\ &= \|t\|^2 \Big\| \sum_i x_i \dot{\otimes} y_i \Big\|^2, \end{aligned} \tag{4.6}$$

the inequality coming from the second part of Lemma 4.2 and the complete positivity of ϕ. Thus $t \otimes \iota$ gives rise to a well-defined, bounded map on $E \otimes_A F$, which extends by continuity to $E \otimes_\phi F$. This extended map is sometimes denoted by $t \otimes \iota$ and sometimes by $\phi_*(t)$. It belongs to $\mathcal{L}(E \otimes_\phi F)$ since it evidently has an adjoint, namely $t^* \otimes \iota$. The map $\phi_*: t \mapsto \phi_*(t)$ is a unital $*$-homomorphism from $\mathcal{L}(E)$ into $\mathcal{L}(E \otimes_\phi F)$ which is strictly continuous on the unit ball of $\mathcal{L}(E)$ (check this!). An equivalent way of saying this is that $\phi_* \in \mathrm{Mor}(\mathcal{K}(E), \mathcal{K}(E \otimes_\phi F))$. It is easy to see that if ϕ is injective then so is ϕ_*.

The fact that $\phi_* \in \mathrm{Mor}(\mathcal{K}(E), \mathcal{K}(E \otimes_\phi F))$ does not of course imply that $\phi_*(\mathcal{K}(E))$ is actually contained in $\mathcal{K}(E \otimes_\phi F)$. However, there is a special case in which this property does hold, namely when $\phi(A)$ is contained in $\mathcal{K}(F)$. We prove this in the following proposition. (Note that Proposition 4.7 may fail if we merely assume that $\phi \in \mathrm{Mor}(A, \mathcal{K}(F))$. For a counterexample, let $E = A = \mathbb{C}$, let B be a nonunital C*-algebra and let $F = B$. The canonical embedding ι of $\mathbb{C}$ as multiples of the identity in $M(B)$ is an element of $\mathrm{Mor}(\mathbb{C}, B)$. Under the canonical unitary equivalences $\mathcal{K}(\mathbb{C}) \cong \mathbb{C}$ and $\mathbb{C} \otimes_\iota B \cong B$, ι_* is also the embedding of $\mathbb{C}$ in $M(B)$, and so $\iota_*(\mathbb{C})$ is not contained in B.) First we need a little lemma.

LEMMA 4.6. *Let E be a Hilbert A-module, F a Hilbert B-module and $\phi: A \to \mathcal{L}_B(F)$ a $*$-homomorphism. For x in E, the equation $\theta_x y = x \dot{\otimes} y$ ($y \in F$) defines an element θ_x of $\mathcal{L}_B(F, E \otimes_\phi F)$ which satisfies*

$$\|\theta_x\| = \|\phi(|x|)\| \leqslant \|x\|,$$
$$\theta_x^* \cdot u \dot{\otimes} v = \phi(\langle x, u\rangle) v \qquad (u \in E,\ v \in F).$$

Proof. Straightforward verification. (Do it!)

PROPOSITION 4.7. *Let E be a Hilbert A-module, F a Hilbert B-module and $\phi: A \to \mathcal{K}_B(F)$ a $*$-homomorphism. Then $\phi_*(\mathcal{K}(E)) \subseteq \mathcal{K}_B(E \otimes_\phi F)$. If ϕ is injective then so is ϕ_*; and if ϕ is surjective then so is ϕ_*.*

Proof. For a in A, x, y, u in E and v in F, we have

$$\begin{aligned} \theta_x \phi(a) \theta_y^* \cdot u \dot{\otimes} v &= \theta_x \phi(a) \phi(\langle y, u\rangle) v \\ &= x \dot{\otimes} \phi(a \langle y, u\rangle) v \\ &= x a \langle y, u\rangle \dot{\otimes} v \\ &= (\theta_{xa,y} u) \dot{\otimes} v \\ &= \phi_*(\theta_{xa,y}) \cdot u \dot{\otimes} v, \end{aligned} \tag{4.7}$$

where θ_x, θ_y are as in Lemma 4.6. Thus

$$\phi_*(\theta_{xa,y}) = \theta_x \phi(a) \theta_y^*, \tag{4.8}$$

which is in $\mathcal{K}_B(E \otimes_\phi F)$ since $\phi(a) \in \mathcal{K}_B(F)$. If (e_i) is an approximate unit for A then $\theta_{xe_i,y} \xrightarrow{i} \theta_{x,y}$ in $\mathcal{K}_A(E)$. Hence $\phi_*(\theta_{xe_i,y}) \xrightarrow{i} \phi_*(\theta_{x,y})$ in the $\mathcal{L}_B(E\otimes_\phi F)$ norm, and it follows that $\phi_*(\theta_{x,y}) \in \mathcal{K}_B(E\otimes_\phi F)$. (Note in passing that if ϕ is nondegenerate then $\phi(e_i) \xrightarrow{i} 1$ strictly, so it follows from (4.8) that $\phi_*(\theta_{x,y}) = \theta_x\theta_y^*$.) Since $\mathcal{K}_A(E)$ is generated by elements of the form $\theta_{x,y}$, it follows that $\phi_*\big(\mathcal{K}_A(E)\big) \subseteq \mathcal{K}_B(E\otimes_\phi F)$.

If ϕ is injective then ϕ_* is always injective, as already noted.

Finally, suppose that ϕ is surjective. Then for z, w in F we can find a in A such that $\phi(a) = \theta_{z,w}$. With x, y, u, v as before, we have

$$\begin{aligned}
\theta_{x\dot\otimes z,y\dot\otimes w}\, u\dot\otimes v &= x\dot\otimes z\langle y\dot\otimes w, u\dot\otimes v\rangle \\
&= x\dot\otimes z\big\langle w, \phi\big(\langle y,u\rangle\big)v\big\rangle \\
&= x\dot\otimes\big(\theta_{z,w}\phi\big(\langle y,u\rangle\big)v\big) \\
&= x\dot\otimes\phi\big(a\langle y,u\rangle\big)v \\
&= \phi_*(\theta_{xa,y})\cdot u\dot\otimes v
\end{aligned}$$

by (4.7). Hence $\theta_{x\dot\otimes z,y\dot\otimes w} = \phi_*(\theta_{xa,y})$. Since $\operatorname{ran}(\phi_*)$ is closed and the C*-algebra $\mathcal{K}_B(E\otimes_\phi F)$ is linearly generated by elements of the form $\theta_{x\dot\otimes z,y\dot\otimes w}$, it follows that ϕ_* is surjective, as required.

References for Chapter 4: [Kas 2], [JenTho].

Chapter 5

The KSGNS construction

This chapter will be about completely positive mappings between C*-algebras, and we begin by establishing some elementary properties of these mappings. What follows can mostly be found in Chapter 3 of [Pau], but we give a self-contained treatment because we want to cover the case of nonunital mappings.

LEMMA 5.1. *If $\rho: A \to B$ is a positive linear map between C*-algebras then ρ is bounded.*

Proof. The proof is essentially the same as the proof that positive linear functionals on C*-algebras are bounded (2.1.8 in [Dix 2], or Theorem 3.3.1 in [Mur]). We first show that ρ is bounded on A_1^+, the positive part of the unit ball of A. If not, then we can find a sequence (a_n) in A_1^+ with $\|\rho(a_n)\| \geqslant n^3$. Then $\sum_{n=1}^{\infty} a_n/n^2$ converges to an element c of A. For each n, we have $c \geqslant a_n/n^2$ and so $\rho(c) \geqslant \rho(a_n)/n^2$. Thus $\|\rho(c)\| \geqslant \|\rho(a_n)/n^2\| \geqslant n$ for all n, which is impossible.

Thus there exists $M > 0$ such that $\|\rho(a)\| \leqslant M\|a\|$ for all a in A^+, from which it is easily seen that ρ is bounded, with $\|\rho\| \leqslant 4M$.

LEMMA 5.2. *Let H be a Hilbert space and suppose that $p, q, t \in \mathcal{L}(H)$, with $p \geqslant 0$ and $q \geqslant 0$. Then*

(i) $\|t\| \leqslant 1$ *if and only if* $\begin{pmatrix} 1 & t \\ t^* & 1 \end{pmatrix} \geqslant 0$ *in* $\mathcal{L}(H \oplus H)$;

(ii) $\begin{pmatrix} p & t \\ t^* & q \end{pmatrix} \geqslant 0$ *if and only if* $|\langle \xi, t\eta \rangle|^2 \leqslant \langle \xi, p\xi \rangle \langle \eta, q\eta \rangle \quad (\xi, \eta \in H)$;

(iii) *if* $\begin{pmatrix} p & t \\ t^* & q \end{pmatrix} \geqslant 0$ *then* $t^*t \leqslant \|p\| q$.

Proof. (ii): For ξ, η in H,

$$\left\langle \begin{pmatrix} \xi \\ \eta \end{pmatrix}, \begin{pmatrix} p & t \\ t^* & q \end{pmatrix} \begin{pmatrix} \xi \\ \eta \end{pmatrix} \right\rangle = \langle \xi, p\xi \rangle + 2\,\mathrm{re}\,\langle \xi, t\eta \rangle + \langle \eta, q\eta \rangle. \tag{5.1}$$

As in the standard proof of the Cauchy–Schwarz inequality, replace η in (5.1) by $\lambda\theta\eta$, where $\lambda \in \mathbf{R}$, and $\theta \in \mathbf{C}$ is chosen with $|\theta| = 1$ to make $\langle \xi, t\eta \rangle$ real. Then the right-hand side of (5.1) is a quadratic form in λ, and the condition for it to be positive semi-definite is the inequality in (ii).

(i) follows easily from (ii), on setting $p = q = 1$.

(iii): From (ii), we have

$$|\langle \xi, t\eta \rangle|^2 \leqslant \langle \xi, p\xi \rangle \langle \eta, q\eta \rangle \qquad (\xi, \eta \in H).$$

Take the supremum over ξ in the unit ball of H to get

$$\langle \eta, t^* t\eta \rangle = \|t\eta\|^2 \leqslant \|p\| \langle \eta, q\eta \rangle,$$

from which (iii) follows.

Lemma 5.3. *Let $\rho: A \to B$ be a completely positive linear map between C^*-algebras, and let (e_i) be an approximate unit for A. Then*

(i) $\|\rho\| = \sup_i \|\rho(e_i)\|$;

(ii) $\rho^{(n)}(a^*)\rho^{(n)}(a) \leqslant \|\rho\| \rho^{(n)}(a^* a) \quad (a \in M_n(A))$.

Proof. Let $M = \sup_i \|\rho(e_i)\|$. Clearly $M \leqslant \|\rho\|$ (which is finite by Lemma 5.1). For a in A, we have

$$\begin{pmatrix} e_i^2 & e_i a \\ a^* e_i & a^* a \end{pmatrix} = \begin{pmatrix} e_i & a \\ 0 & 0 \end{pmatrix}^* \begin{pmatrix} e_i & a \\ 0 & 0 \end{pmatrix} \geqslant 0.$$

Hence

$$\begin{pmatrix} \rho(e_i^2) & \rho(e_i a) \\ \rho(a^* e_i) & \rho(a^* a) \end{pmatrix} \geqslant 0,$$

and by Lemma 5.2(iii) we have

$$\begin{aligned} \rho(a^* e_i)\rho(e_i a) &\leqslant \|\rho(e_i^2)\|\, \rho(a^* a) \\ &\leqslant \|\rho(e_i)\|\, \rho(a^* a) \end{aligned}$$

since $e_i^2 \leqslant e_i$. Taking the limit along (e_i) (and recalling that ρ is continuous), we have

$$\rho(a^*)\rho(a) \leqslant M\rho(a^*a).$$

Thus if $\|a\| \leqslant 1$ then we have

$$\|\rho(a)\|^2 \leqslant M\|\rho(a^*a)\| \leqslant M\|\rho\|.$$

Taking the supremum over a in the unit ball of A, we have $\|\rho\|^2 \leqslant M\|\rho\|$, and hence $\|\rho\| \leqslant M$. So we have proved (i), and also (ii) in the case $n = 1$.

Now observe that for $n \geqslant 1$, if $e_i^{(n)} = e_i \otimes 1 \in A \otimes M_n(\mathbb{C}) \cong M_n(A)$ then $(e_i^{(n)})$ is an approximate unit for $M_n(A)$, and $\|\rho^{(n)}(e_i^{(n)})\| = \|\rho(e_i)\|$. Applying what we have already proved to the completely positive map $\rho^{(n)}$, we see that (ii) holds in general.

The proof of the above Schwarz inequality also shows that $\|\rho^{(n)}\| = \|\rho\|$ for all n. Thus ρ is completely bounded, with $\|\rho\|_{\mathrm{cb}} = \|\rho\|$ (where by definition $\|\rho\|_{\mathrm{cb}} = \sup_n \|\rho^{(n)}\|$).

LEMMA 5.4. *Let $\rho: A \to B$ be a completely positive mapping between C*-algebras, and let $a_1, \ldots, a_n \in A$. Then $(\rho(a_i^*)\rho(a_j)) \leqslant \|\rho\|(\rho(a_i^*a_j))$ in $M_n(B)$.*

Proof. Apply Lemma 5.3(ii) to the element

$$\begin{pmatrix} a_1 & a_2 & \ldots & a_n \\ 0 & 0 & \ldots & 0 \\ \vdots & \vdots & \ddots & \vdots \\ 0 & 0 & \ldots & 0 \end{pmatrix}$$

of $M_n(A)$.

That concludes our study of elementary properties of completely positive maps, and we now move on to consider completely positive maps in the context of Hilbert C*-modules. The key question that we ask is: Can we perform the interior tensor product construction described in the previous chapter, but using a completely positive mapping ρ in place of the $*$-homomorphism ϕ?

Let E be a Hilbert A-module, F a Hilbert B-module and $\rho: A \to \mathcal{L}(F)$ a completely positive map. Even though ρ need not be a $*$-homomorphism, we can still try to form $E \otimes_\rho F$. If we attempt to follow the previous construction in all details then we shall run into trouble almost as soon as we start, because this time F need not be a left A-module under the action $(a, y) \mapsto \rho(a)y$. So we cannot hope to form the algebra tensor product $E \otimes_A F$ or to identify $E \otimes_\rho F$ with its completion. However, we can still define a sesquilinear B-valued linear form on the right B-module $E \otimes_{\text{alg}} F$, given on simple tensors by

$$\langle x_1 \otimes y_1, x_2 \otimes y_2 \rangle = \langle y_1, \rho(\langle x_1, x_2 \rangle) y_2 \rangle \qquad (x_1, x_2 \in E,\ y_1, y_2 \in F).$$

The argument in the first paragraph of the proof of Proposition 4.5 still applies, and shows that $E \otimes_{\text{alg}} F$ becomes a semi-inner-product B-module in this way. We define $E \otimes_\rho F$ to be the Hilbert B-module obtained by completing the quotient inner-product B-module $(E \otimes_{\text{alg}} F)/N_\rho$, where

$$N_\rho = \{z \in E \otimes_{\text{alg}} F : \langle z, z \rangle = 0\}.$$

We shall again use the notation $x \dot{\otimes} y$ for the coset $x \otimes y + N_\rho$, considered as an element of $E \otimes_\rho F$.

Furthermore, the calculation (4.6) is still valid with ρ in place of ϕ, and shows that, for t in $\mathcal{L}(E)$, if $\rho_*(t)$ is defined on simple tensors by

$$\rho_*(t) \cdot x \dot{\otimes} y = tx \dot{\otimes} y \qquad (x \in E,\ y \in F),$$

then $\rho_*(t)$ extends to an element (still denoted by $\rho_*(t)$) of $\mathcal{L}(E \otimes_\rho F)$. Also, just as previously, the map $\rho_*: t \to \rho_*(t)$ is a unital $*$-homomorphism from $\mathcal{L}(E)$ to $\mathcal{L}(E \otimes_\rho F)$ which is strictly continuous on the unit ball.

For the remainder of this chapter we shall be interested only in the case $E = A$, and we shall consider a completely positive map $\rho: A \to \mathcal{L}(F)$, where F is a Hilbert B-module. To make further progress, we shall need to impose some condition of nondegeneracy on ρ. In the case of a $*$-homomorphism, we saw in Proposition 2.5 that there are three conditions, all equivalent, that could be taken as the definition of a nondegenerate $*$-homomorphism. For completely positive mappings, we shall investigate later in this chapter the extent to which the analogous three conditions are still equivalent. For now, however, we take condition (iii) of Proposition 2.5 as the definition of nondegeneracy for completely positive mappings. Namely, $\rho: A \to \mathcal{L}(F)$ is nondegenerate if $\rho(e_i) \to 1$ strictly in $\mathcal{L}(F)$, for some approximate unit (e_i) of A.

In fact, it will be convenient to consider a weaker condition than nondegeneracy. Given a completely positive map $\rho: A \to \mathcal{L}(F)$, we say that ρ is *strict* if $(\rho(e_i))$ is strictly Cauchy in $\mathcal{L}(F)$, for some approximate unit (e_i) of A. Clearly nondegeneracy implies strictness. If A is unital then the condition of strictness is automatically satisfied (whereas ρ is nondegenerate only if ρ is unital); but when A is nonunital, strictness imposes a genuine limitation on ρ (for example, the inclusion map $C_0((0,1]) \to C([0,1])$ is not strict). Another important situation where strictness is automatic is when $B = \mathbb{C}$: for then F is a Hilbert space H, and $(\rho(e_i))$ is a bounded increasing directed net of positive elements in $\mathcal{L}(H)$, which must have a strong (and therefore strict) limit in $\mathcal{L}(H)$ (Theorem 4.2.2 in [Mur] or Lemma 5.1.4 of [KadRin]).

It is easy to check that the unit ball of $\mathcal{L}(F)$ is complete for the strict topology. (Exercise: prove this, using the proof of [KadRin], Proposition 2.5.11 as a guide if needed.) So $\rho: A \to \mathcal{L}(F)$ is strict if and only if there is a positive element p in the ball of radius $\|\rho\|$ in $\mathcal{L}(F)$ such that $\rho(e_i) \to p$ strictly.

PROPOSITION 5.5. *Suppose that A, B are C^*-algebras, E, F are Hilbert B-modules, $\pi: A \to \mathcal{L}(E)$ is a nondegenerate $*$-homomorphism and $v \in \mathcal{L}(F, E)$. Then $\rho: A \to \mathcal{L}(F)$ defined by*

$$\rho(a) = v^* \pi(a) v \qquad (a \in A) \tag{5.2}$$

is a strict completely positive mapping.

Proof. It is routine (cf. Exercise 11.5.15 in [KadRin]) to check that ρ is completely positive. If (e_i) is an approximate unit for A then $\pi(e_i) \to 1$ strictly in $\mathcal{L}(E)$, since π is nondegenerate, and therefore $\rho(e_i) \to v^* v$ strictly in $\mathcal{L}(F)$. Hence ρ is strict.

Our main goal in this chapter will be to prove a converse to Proposition 5.5, stating that every strict completely positive map $\rho: A \to \mathcal{L}(F)$ is given by a formula like (5.2).

For a nondegenerate $*$-homomorphism $\phi: A \to \mathcal{L}(F)$, we saw in the previous chapter that $A \otimes_\phi F$ is unitarily equivalent to F. In the case of a completely positive map $\rho: A \to \mathcal{L}(F)$, the calculation analogous to (4.5) goes as follows, where $a_1, \ldots, a_n \in A$, $y_1, \ldots, y_n \in F$:

$$\begin{aligned} \Big\|\sum_i \rho(a_i)y_i\Big\|^2 &= \Big\|\sum_{i,j}\langle y_i, \rho(a_i^*)\rho(a_j)y_j\rangle\Big\| \\ &\leqslant \|\rho\|\Big\|\sum_{i,j}\langle y_i, \rho(a_i^* a_j)y_j\rangle\Big\| \\ &= \|\rho\|\Big\|\sum_i a_i \dot\otimes y_i\Big\|^2, \end{aligned} \tag{5.3}$$

the inequality coming from Lemma 5.4. Thus the map $a\dot\otimes y \mapsto \rho(a)y$ extends by linearity and continuity to a bounded map $w: A\otimes_\rho F \to F$ satisfying $\|w\| \leqslant \|\rho\|^{\frac{1}{2}}$ and

$$\overline{\operatorname{ran}(w)} = \overline{\rho(A)F}. \tag{5.4}$$

Now assume that ρ is strict, with $\rho(e_i) \to p \in \mathcal{L}(F)$ strictly, where $(e_i)_{i\in I}$ is an approximate unit for A. We denote the ordering on the index set I by $<$. For each i in I, define $v_i: F \to A\otimes_\rho F$ by $v_i(y) = e_i\dot\otimes y$. For $i > j$ in I and y in F,

$$\begin{aligned} \|v_i(y) - v_j(y)\|^2 &= \|\langle y, \rho((e_i - e_j)^2)y\rangle\| \\ &\leqslant \|\langle y, \rho(e_i - e_j)y\rangle\| \end{aligned}$$

and, since $(\rho(e_i))$ is strictly Cauchy, this can be made arbitrarily small provided i, j are "large enough" in I. Thus $\big(v_i(y)\big)$ is Cauchy in $A\otimes_\rho F$ and hence converges to an element $v(y)$. For a in A and y, z in F, we have

$$\begin{aligned} \langle a\dot\otimes y, vz\rangle &= \lim_i \langle a\dot\otimes y, e_i\dot\otimes z\rangle \\ &= \lim_i \langle y, \rho(a^* e_i)z\rangle \\ &= \langle y, \rho(a^*)z\rangle = \langle w(a\dot\otimes y), z\rangle. \end{aligned}$$

By linearity and continuity, $\langle x, vz\rangle = \langle wx, z\rangle$ for all x in $A\otimes_\rho F$ and therefore $v = w^*$. Thus $v \in \mathcal{L}(F, A\otimes_\rho F)$, $w \in \mathcal{L}(A\otimes_\rho F, F)$.

Associated with ρ there is (as noted earlier in the chapter) a unital $*$-homomorphism $\rho_*: M(A) \to \mathcal{L}(A\otimes_\rho F)$ which is strictly continuous on the unit ball and is given by

$$\rho_*(c)\cdot a\dot\otimes y = ca\dot\otimes y \qquad (c \in M(A),\ a \in A,\ y \in F).$$

By Proposition 2.5, the restriction of ρ_* to A is nondegenerate, so that $\rho_*(A) \cdot A \otimes_\rho F$ is dense in $A \otimes_\rho F$. In fact, ρ_* satisfies a slightly stronger nondegeneracy condition, namely that $\rho_*(A)v(F)$ is dense in $A \otimes_\rho F$. For if $a \in A$ and $y \in F$ then

$$\begin{aligned}\|\rho_*(a) \cdot e_i \dot{\otimes} y - a \dot{\otimes} y\|^2 &= \|(ae_i - a)\dot{\otimes} y\|^2 \\ &= \left\|\langle y, \rho((1-e_i)a^*a(1-e_i))y\rangle\right\| \xrightarrow{i} 0,\end{aligned}$$

so that $\rho_*(a) \cdot vy = a \dot{\otimes} y$. Hence (recall our convention that the product of two sets means the linear span of products of their elements) $\rho_*(A)vF$ contains all finite sums of elements of the form $a \dot{\otimes} y$ and is therefore dense in $A \otimes_\rho F$.

We now change notation, and write F_ρ for the Hilbert B-module $A \otimes_\rho F$, π_ρ for the $*$-homomorphism $\rho_*: M(A) \to \mathcal{L}(F_\rho)$ and v_ρ for the operator v in $\mathcal{L}(F, F_\rho)$. For a in A and y in F, we have

$$\begin{aligned}v_\rho^* \pi_\rho(a) v_\rho y &= \lim_i v_\rho^* \pi_\rho(a) \cdot e_i \dot{\otimes} y \\ &= \lim_i v_\rho^* \cdot ae_i \dot{\otimes} y \\ &= \lim_i \rho(ae_i)y = \rho(a)y,\end{aligned}$$

and therefore $\rho(a) = v_\rho^* \pi_\rho(a) v_\rho$. We have proved the first part of the following fundamental theorem.

THEOREM 5.6. *Let A, B be C^*-algebras, let F be a Hilbert B-module and let $\rho: A \to \mathcal{L}(F)$ be a strict completely positive mapping.*

(i) *There exist a Hilbert B-module F_ρ, a $*$-homomorphism $\pi_\rho: A \to \mathcal{L}(F_\rho)$ and an element v_ρ of $\mathcal{L}(F, F_\rho)$, such that*

$$\rho(a) = v_\rho^* \pi_\rho(a) v_\rho \qquad (a \in A), \tag{5.5}$$

$$\pi_\rho(A) v_\rho F \ \textit{ is dense in } F_\rho. \tag{5.6}$$

(ii) *If G is a Hilbert B-module, $\pi: A \to \mathcal{L}(G)$ is a $*$-homomorphism, $w \in \mathcal{L}(F, G)$ and*

$$\rho(a) = w^* \pi(a) w \qquad (a \in A), \tag{5.7}$$

$$\pi(A) w F \ \textit{ is dense in } G, \tag{5.8}$$

then there is a unitary u in $\mathcal{L}(F_\rho, G)$ such that

$$\pi(a) = u\pi_\rho(a)u^* \qquad (a \in A)$$

and $w = uv_\rho$.

Proof. It only remains to establish the uniqueness assertion (ii). If $a_1, \ldots, a_n \in A$ and $y_1, \ldots, y_n \in F$ then

$$\begin{aligned}\Big|\sum_i \pi(a_i)wy_i\Big|^2 &= \sum_{i,j}\langle y_i, w^*\pi(a_i^*a_j)wy_j\rangle \\ &= \sum_{i,j}\langle y_i, \rho(a_i^*a_j)y_j\rangle \\ &= \sum_{i,j}\langle y_i, v_\rho^*\pi_\rho(a_i^*a_j)v_\rho y_j\rangle = \Big|\sum_i \pi_\rho(a_i)v_\rho y_i\Big|^2,\end{aligned}$$

by (5.7) and (5.5). Thus the map $u: \sum \pi_\rho(a_i)v_\rho y_i \mapsto \sum \pi(a_i)wy_i$ is (well-defined and) isometric, and extends by continuity to a unitary map u from F_ρ onto G (because of the nondegeneracy conditions (5.6) and (5.8)). Since

$$\begin{aligned}u\pi_\rho(a)u^* \sum \pi(a_i)wy_i &= u\pi_\rho(a)\sum \pi_\rho(a_i)v_\rho y_i \\ &= u\sum \pi_\rho(aa_i)v_\rho y_i \\ &= \sum \pi(aa_i)wy_i = \pi(a)\sum \pi(a_i)wy_i,\end{aligned}$$

we have $u\pi_\rho(a)u^* = \pi(a)$ $(a \in A)$. Also, from the definition of u,

$$u\pi_\rho(e_i)v_\rho y = \pi(e_i)wy \qquad (y \in F).$$

Since π_ρ and π are nondegenerate, both $\pi_\rho(e_i)$ and $\pi(e_i)$ converge strictly to the identity as we take the limit over (e_i). Thus $uv_\rho y = wy$, and hence $uv_\rho = w$.

The unique (up to unitary equivalence) triple $(F_\rho, \pi_\rho, v_\rho)$ obtained from ρ as in Theorem 5.6 will be called the KSGNS construction associated with ρ (for Kasparov, Stinespring, Gel′fand, Naĭmark, Segal). If $F = B = \mathbf{C}$ then the KSGNS construction reduces to the classical GNS construction. If $B = \mathbf{C}$ (so that F is a Hilbert space) then we get the Stinespring construction (Exercise 11.5.17 in [KadRin]). In the context of Hilbert C*-modules, the construction was given by Kasparov [Kas 1].

COROLLARY 5.7. *Let A, B be C^*-algebras, let F be a Hilbert B-module and let $\rho: A \to \mathcal{L}(F)$ be completely positive. Then ρ is strict if and only if there is a completely positive map $\bar{\rho}: M(A) \to \mathcal{L}(F)$, strictly continuous on the unit ball, whose restriction to A is equal to ρ. Also, ρ is nondegenerate if and only if $\bar{\rho}$ is unital.*

Proof. If $\bar{\rho}$ exists, and (e_i) is an approximate unit for A, then $\rho(e_i) \to \bar{\rho}(1)$ strictly, from which it follows that ρ is strict. For the converse implication, suppose that ρ is strict. Then Theorem 5.6 applies, and gives a $*$-homomorphism $\pi_\rho: A \to \mathcal{L}(F_\rho)$ which is strictly continuous on the unit ball and therefore, by the results in Chapter 2, has an extension $\bar{\pi}_\rho: M(A) \to \mathcal{L}(F_\rho)$. Define

$$\bar{\rho}(c) = v_\rho^* \bar{\pi}_\rho(c) v_\rho \qquad (c \in M(A)). \tag{5.9}$$

Then $\bar{\rho}$ evidently has the desired properties. Also, since $\rho(e_i) \to \bar{\rho}(1)$ it is clear that ρ is nondegenerate if and only if $\bar{\rho}$ is unital.

We can now clarify the notion of nondegeneracy, as promised earlier in the chapter. By analogy with Proposition 2.5, there are three possible definitions of nondegeneracy for a completely positive map $\rho: A \to \mathcal{L}(F)$, namely

(i) $\rho(A)F$ is dense in F;

(ii) ρ is the restriction to A of a unital completely positive map $\bar{\rho}: M(A) \to \mathcal{L}(F)$ which is strictly continuous on the unit ball;

(iii) for some approximate unit (e_i) of A, $\rho(e_i) \to 1$ strictly.

We decided to take (iii) as the definition of nondegeneracy, and we have just seen in Corollary 5.7 that (ii) is equivalent to (iii). Clearly (iii) implies (i). But (i) does not imply (iii), as the following example shows. In fact, condition (i) does not appear to be useful in the context of completely positive maps.

For the example, take $B = F = C([0,1])$, let $A = \{f \in B: f(0) = 0\}$ and define $\rho: A \to B$ by

$$\rho(f)(\lambda) = \tfrac{1}{2}\big(f(\lambda) + f(1-\lambda)\big) \qquad (0 \leqslant \lambda \leqslant 1).$$

It is left to the reader to check that ρ is completely positive and that $\rho(A)B = B$, so that ρ satisfies (i). But ρ does not satisfy (iii). In fact, if (e_i) is an approximate identity for A then the strict limit of $(\rho(e_i))$ (if it existed) would have to be a function taking the value 1 on $(0,1)$ and taking the value $\frac{1}{2}$ at $\lambda = 0$ and $\lambda = 1$. This is clearly contradictory since functions in B have to be continuous.

The significance of nondegeneracy for completely positive mappings lies in the following fact. If $\rho: A \to \mathcal{L}(F)$ is nondegenerate then by Corollary 5.7 $\bar{\rho}$ is unital, and so by (5.9) $v_\rho^* v_\rho = 1$. Thus v_ρ is an isometry, which means that $v_\rho v_\rho^*$ is a projection, and F can be identified with the complemented submodule $\operatorname{ran}(v_\rho v_\rho^*)$ of F_ρ. In this way, ρ is identified with the compression of the nondegenerate $*$-homomorphism π_ρ to a complemented submodule of F_ρ.

Another noteworthy special case of the KSGNS construction occurs when v_ρ^*, rather than v_ρ, is an isometry. This happens when ρ is a $*$-homomorphism. To see this, suppose that $\phi: A \to \mathcal{L}(F)$ is a $*$-homomorphism which is strict but not necessarily nondegenerate. Then the inequality in (5.3) becomes an equality (and of course $\|\phi\| = 1$, unless ϕ is the zero mapping, which we exclude from consideration). It follows that $w = v_\rho^*$ is an isometry. We use this to prove the following result, which is essentially the same as Proposition 1.1.13 in [JenTho] and should be compared with Proposition 2.5.

PROPOSITION 5.8. *Let A, B be C^*-algebras and let F be a Hilbert B-module. For a $*$-homomorphism $\phi: A \to \mathcal{L}(F)$, the following conditions are equivalent:*

(i) *$\overline{\phi(A)F}$ is a complemented submodule of F;*

(ii) *ϕ is the restriction to A of a $*$-homomorphism $\bar{\phi}: M(A) \to \mathcal{L}(F)$ which is strictly continuous on the unit ball;*

(iii) *ϕ is strict.*

If these conditions hold then $\bar{\phi}(1)$, which is the strict limit of $(\phi(e_i))$ for an approximate unit (e_i) of A, is the projection from F onto $\overline{\phi(A)F}$.

Proof. (i) $\Rightarrow$ (ii): Let p be the projection from F onto $F_0 = \overline{\phi(A)F}$. Then $p \in \mathcal{L}(F, F_0)$ and p^* is the embedding of F_0 in F. For a in A, define $\tilde{\phi}(a) = p\phi(a)p^*$. Then $\tilde{\phi}: A \to \mathcal{L}(F_0)$ is a nondegenerate $*$-homomorphism. Reason: just as in the case of Hilbert space operators, one can see that $\phi(A)F_0^\perp = \{0\}$, from which it easily follows that $\tilde{\phi}$ is multiplicative and nondegenerate. From the results in Chapter 2, $\tilde{\phi}$ has an extension (which we still call $\tilde{\phi}$) to a $*$-homomorphism from $M(A)$ to $\mathcal{L}(F_0)$ which is strictly continuous on the unit ball. Define $\bar{\phi}(c) = p^*\tilde{\phi}(c)p \quad (c \in M(A))$.

(ii) $\Rightarrow$ (iii) follows from Corollary 5.7.

(iii) $\Rightarrow$ (i): By the remarks preceding the proposition, v_ϕ^* is an isometry, whose range (by (5.4)) is $\overline{\phi(A)F}$. Thus F_0 is the range of $v_\phi^* v_\phi$ and is therefore complemented.

The last assertion of the proposition is left as an (easy) exercise.

For a completely positive mapping that is not strict, it is still possible to find a representation of the form (5.2), but in this case one must abandon the requirement that π should be nondegenerate. To see this, let $\rho: A \to \mathcal{L}(F)$ be completely positive, and let A^+ be the C*-algebra obtained by adjoining an identity to A. Using the elementary C*-algebraic fact that if $a \geqslant 0$ in A^+ then $a \leqslant 1$ if and only if $\|a\| \leqslant 1$, together with Lemma 5.3, one sees that if we define $\rho(1) = \|\rho\|1$ then ρ becomes a completely positive map from A^+ to $\mathcal{L}(F)$, which is necessarily strict since A^+ is unital. Thus $\rho(a) = v_\rho^* \pi_\rho(a) v_\rho$ for all a in A^+, and in particular for all a in A. Note that π_ρ is nondegenerate as a $*$-homomorphism from A^+ to $\mathcal{L}(F)$, but that its restriction to A need not be. For non-strict mappings, one must also abandon the uniqueness assertion of Theorem 5.6.

We conclude this chapter by looking at a class of completely positive mappings that we call retractions. These are a mild generalisation of the conditional expectations that we met in Chapter 1.

Let A be a C*-algebra and let B be a strictly closed C*-subalgebra of $M(A)$. A *retraction* from A to B is a positive linear map $\psi: A \to B$ satisfying the conditions

(i) $\psi(ab) = \psi(a)b \quad (a \in A,\ b \in B)$,

(ii) $\psi(A)$ is strictly dense in B,

(iii) for some approximate unit (e_i) of A, $\big(\psi(e_i)\big)$ converges strictly in $M(A)$ to a projection $p \in B$.

If A is unital (and $p = 1$) then a retraction is just the same as a conditional expectation. But in the nonunital case the above definition is flexible enough to include some interesting examples. For instance, any state of A

is a retraction from A to $\mathbf{C}$ (identified with the subalgebra $\mathbf{C}1$ of $M(A)$). Also, if p is any projection in $M(A)$ then the compression map $a \mapsto pap$ is a retraction from A to the corner $pM(A)p$ of $M(A)$.

LEMMA 5.9. *If $\psi: A \to B \subseteq M(A)$ is a retraction then ψ is completely positive.*

Proof. By the argument in the proof of Lemma 4.3(i), it suffices to prove that $(\psi(a_i^* a_j)) \geqslant 0$ in $M_n(B)$ for all $a_1, \dots, a_n$ in A. Identifying $M_n(B)$ with $\mathcal{K}(B^n)$, we see from Lemma 4.1 that this will hold if

$$\sum_{i,j=1}^{n} b_i^* \psi(a_i^* a_j) b_j \geqslant 0 \qquad (b_1, \dots, b_n \in B). \tag{5.10}$$

But by condition (i) above (and its adjoint), the left-hand side of (5.10) is equal to $\psi(c^* c)$, where $c = \sum_i a_i b_i$, and this is positive since ψ is a positive mapping.

PROPOSITION 5.10. *Let A be a C^*-algebra and let B be a strictly closed C^*-subalgebra of $M(A)$. A linear map $\psi: A \to B$ is a retraction if and only if ψ is the restriction to A of an idempotent contraction $\bar{\psi}: M(A) \to B$ which is strictly continuous on the unit ball.*

Proof. Suppose that $\psi: A \to B$ is a retraction. Since ψ, regarded as a map from A to $M(A)$, is completely positive (by Lemma 5.9) and strict (by (iii)), it follows from Corollary 5.7 that ψ has an extension $\bar{\psi}: M(A) \to M(A)$ which is strictly continuous on the unit ball and is a contraction since $\|\bar{\psi}\| = \|\bar{\psi}(1)\| = \|p\| = 1$. By strict continuity and the fact that B is strictly closed, we see that the range of $\bar{\psi}$ is contained in B. Also by strict continuity, it follows from (i) that

$$\bar{\psi}(cb) = \bar{\psi}(c)b \qquad (c \in M(A),\ b \in B). \tag{5.11}$$

With $c = 1$, this equation and its adjoint give

$$\bar{\psi}(b) = pbp \qquad (b \in B), \tag{5.12}$$

and in particular $\bar{\psi}(p) = p$.

For any c in $M(A)$ with $0 \leqslant c \leqslant 1$ we have $0 \leqslant \bar{\psi}(c) \leqslant p$ and therefore $\bar{\psi}(c) = p\bar{\psi}(c)p$. By linearity, this holds for all c in $M(A)$:

$$\bar{\psi}(c) = p\bar{\psi}(c)p \qquad (c \in M(A)). \tag{5.13}$$

From (5.12) and (5.13) it follows that, for all a in A,

$$\bar{\psi}(\psi(a)) = p\psi(a)p = \psi(a).$$

By (ii), we have $\bar{\psi}(b) = b$ for all b in B and so $\bar{\psi}$ is idempotent.

Conversely, if $\bar{\psi}: M(A) \to B$ is an idempotent contraction then it follows from Exercise 10.5.86(iv) of [KadRin] that $\bar{\psi}$ satisfies (5.11). It is easy to deduce from this that if $\bar{\psi}$ is strictly continuous on the unit ball then its restriction to A satisfies the conditions for a retraction.

The localisation construction from Chapter 1 can be carried out for retractions. Observe first that if E is a Hilbert A-module then E can be made into a Hilbert $M(A)$-module. This is done as follows. If $c \in M(A)$ and (e_i) is an approximate unit for A then, for x in E,

$$|xe_ic - xe_jc|^2 = c^*(e_i - e_j)|x|^2(e_i - e_j)c \xrightarrow{i,j} 0.$$

So (xe_ic) is Cauchy, with a limit that we call xc. It is straightforward to check that this makes E into a right $M(A)$-module (the action of $M(A)$ on E extending that of A), and that with this action E is a Hilbert $M(A)$-module.

Now suppose that $\psi: A \to B \subseteq M(A)$ is a retraction. We define a B-valued semi-inner product $\langle \cdot, \cdot \rangle_\psi$ on E by

$$\langle x, y \rangle_\psi = \psi(\langle x, y \rangle) \qquad (x, y \in E).$$

This makes E into a semi-inner-product B-module. Factoring out the kernel $N_\psi = \{x \in E: \langle x, x \rangle_\psi = 0\}$ and completing E/N_ψ in the usual way (as described in Chapter 1), we obtain a Hilbert B-module which we denote by E_ψ and which we call the *localisation of E by ψ*.

For t in $\mathcal{L}(E)$ and x in E, we have $0 \leqslant |tx|^2 \leqslant \|t\|^2|x|^2$ (Proposition 1.2), from which it follows that $t(N_\psi) \subseteq N_\psi$. Thus the map

$$\pi_\psi(t): x + N_\psi \mapsto tx + N_\psi$$

is well-defined on E/N_ψ, and bounded, with $\|\pi_\psi(t)\| \leqslant \|t\|$. Extending $\pi_\psi(t)$ by continuity to E_ψ, we obtain an element $\pi_\psi(t)$ of $\mathcal{L}(E_\psi)$. This defines a unital $*$-homomorphism $\pi_\psi: \mathcal{L}(E) \to \mathcal{L}(E_\psi)$, the *localisation of* $\mathcal{L}(E)$ *by* ψ. If ψ is faithful (that is, $\psi(a) = 0 \Rightarrow a = 0$, for $a \geqslant 0$ in A) then π_ψ is an isomorphism.

In particular, if E is a Hilbert A-module then every state ρ of A gives rise to a localisation E_ρ which is a Hilbert space (a Hilbert $\mathbb{C}$-module), and π_ρ is a unital $*$-representation of $\mathcal{L}(E)$ on E_ρ.

For a final example, let E be a Hilbert A-module and n a positive integer. Then E^n can be regarded either as a Hilbert A-module (Chapter 1) or as a Hilbert $M_n(A)$-module (Chapter 4). Write τ for the A-valued trace on $M_n(A)$, defined by

$$\tau\big((a_{ij})\big) = \frac{1}{n}\sum_{i=1}^{n} a_{ii}.$$

Then τ can be regarded as a retraction on $M_n(A)$ provided that we identify $A = \operatorname{ran}(\tau)$ with a subalgebra of $M_n(A)$ via the diagonal embedding

$$a \mapsto \begin{pmatrix} a & 0 & \dots & 0 \\ 0 & a & \dots & 0 \\ \vdots & \vdots & \ddots & \vdots \\ 0 & 0 & \dots & a \end{pmatrix} \qquad (a \in A).$$

It is easy to see that τ is faithful and that the associated localisation map π_τ is the canonical isomorphism from $\mathcal{L}_{M_n(A)}(E^n)$ to $\mathcal{L}_A(E^n)$.

References for Chapter 5: [Kas 1], [JenTho], [Tak].

Chapter 6

Stabilisation or absorption

Given a C*-algebra A, can we hope to classify all Hilbert A-modules up to unitary equivalence? In that generality, the question is too ambitious to have a positive answer. Indeed, if H is any Hilbert space then we can form the Hilbert A-module $H \otimes A$; and if $\dim H$ is allowed to have unrestricted cardinality then a classification even of these modules is going to involve set-theoretic considerations. So it is not surprising that to make progress we need to impose some restrictions on the size of the C*-algebra A and the module E. What we shall prove in this chapter is that, given a mild countability condition on E, the A-module $H \otimes A$ (where H is a separable space) has the universal property that any other Hilbert A-module E occurs as a complemented submodule of it. We begin by examining some countability conditions on A and E.

Recall ([Ped], 3.10.4) that a positive element h of a C*-algebra A is called ***strictly positive*** if $\rho(h) > 0$ for every state ρ of A. The following useful little result is Lemma 1.1.21 of [JenTho].

LEMMA 6.1. *A positive element h of the C*-algebra A is strictly positive if and only if the closed right ideal generated by h is the whole of A.*

Proof. If ρ is a state of A and $\rho(h) = 0$ then by the Cauchy–Schwarz inequality

$$|\rho(ha)|^2 = |\rho(h^{\frac{1}{2}}h^{\frac{1}{2}}a)|^2 \leqslant \rho(h)\rho(a^*ha) = 0 \qquad (a \in A). \tag{6.1}$$

So ρ vanishes on $\overline{hA}$. This cannot happen if $\overline{hA} = A$, and so h must be strictly positive. Conversely, if $\overline{hA} \neq A$ then by 2.9.4 of [Dix 2] or 10.2.9 of

[KadRin] there is a state ρ of A with $\rho(\overline{hA}) = 0$. If (e_i) is an approximate unit for A then $\rho(he_i) = 0$ for all i, hence $\rho(h) = 0$ and h is not strictly positive.

A C*-algebra is called *σ-unital* if it has a countable approximate unit. It is proved in [Ped], Proposition 3.10.5 that a C*-algebra is σ-unital if and only if it has a strictly positive element. A commutative C*-algebra $C_0(X)$ is σ-unital if and only if X is σ-compact.

If E is a Hilbert A-module and $Z \subseteq E$ then we say that Z is a *generating set* for E if the closed submodule of E generated by Z is the whole of E. We say that E is countably generated if it has a countable generating set.

If A is σ-unital then A (considered as a Hilbert A-module) is countably (in fact, singly) generated, for if $h \in A$ is strictly positive then $\{h\}$ is a generating set. Conversely, if A is countably generated, with a generating set $\{a_n : n \geqslant 1\}$, then A is σ-unital: for we may assume that $a_n \geqslant 0$ and $\|a_n\| \leqslant 1$ for all n (why?), and if $h = \sum_n 2^{-n} a_n$ then h is strictly positive, since for any state ρ of A it follows from the Cauchy–Schwarz inequality (6.1) that we must have $\rho(a_n) > 0$ for some n, and hence $\rho(h) > 0$.

Let H be a separable, infinite-dimensional Hilbert space. The Hilbert A-module $H \otimes A$ plays a special role in the theory of Hilbert C*-modules, and in future we shall denote it by H_A. If A is σ-unital then H_A is countably generated, since if $\{\varepsilon_n\}$ is an orthonomal basis for H and h is a strictly positive element of A then $\{\varepsilon_n \otimes h\}$ is a generating set for H_A.

We can now state the first main theorem of this chapter, known as Kasparov's stabilisation theorem, or sometimes as Kasparov's absorption theorem. Intuitively, the idea of the theorem is that H_A is big enough to absorb any countably generated Hilbert A-module; or alternatively that once a module reaches the size of H_A it stabilises, and cannot get any "bigger".

THEOREM 6.2. *If A is a C*-algebra and E is a countably generated Hilbert A-module then $E \oplus H_A \approx H_A$.*

Proof. If A is not unital, let A^+ denote the unitisation of A. Recall from Chapter 1 that E can be regarded as a Hilbert A^+-module; and that $\overline{EA} = E$. It is easy to check that the map

$$(\xi \otimes 1)a \mapsto \xi \otimes a \qquad (\xi \in H,\ a \in A)$$

extends to a unitary equivalence between the Hilbert A-modules $\overline{H_{A^+}A}$ and H_A. Similarly, $\overline{(E \oplus H_{A^+})A} \approx E \oplus H_A$ as Hilbert A-modules. If we can find a unitary A^+-module isomorphism $u: H_{A^+} \to E \oplus H_{A^+}$ then the restriction of u to $\overline{H_{A^+}A}$, together with the above unitary equivalences, will provide a unitary equivalence between H_A and $E \oplus H_A$. So it will be sufficient to prove the theorem in the unital case; and we assume for the rest of the proof that A has identity 1.

Let y_n be a sequence of unit vectors in E in which each element of a generating set occurs infinitely often. Let $e_n = \varepsilon_n \otimes 1$, where $\{\varepsilon_n\}$ is an orthonormal basis for H. Then $\{e_n\}$ is a generating set for H_A. We regard E and H_A as (orthogonal) submodules of $E \oplus H_A$ in the obvious way, and we define $t \in \mathcal{K}(H_A, E \oplus H_A)$ by

$$t = \sum_{n=1}^{\infty} (2^{-n}\theta_{y_n,e_n} + 4^{-n}\theta_{e_n,e_n}),$$

so that $te_n = 2^{-n}y_n + 4^{-n}e_n$. For each m with $y_m = y_n$, we have

$$t(2^m e_m) = y_n + 2^{-m}e_m.$$

Since there are infinitely many such m, it follows that y_n is in the closure of the range of t; and since

$$e_n = t(4^n e_n) - 2^n y_n,$$

it follows that e_n is in $\overline{\mathrm{ran}(t)}$ too. Therefore $\mathrm{ran}(t)$ is dense in $E \oplus H_A$. Also, $t^* = \sum_n (2^{-n}\theta_{e_n,y_n} + 4^{-n}\theta_{e_n,e_n})$ so that $t^*(4^n e_n) = e_n$. Thus $\mathrm{ran}(t^*)$ is dense in H_A. By Proposition 3.8, $E \oplus H_A \approx H_A$.

We say that a closed submodule F of a Hilbert C*-module E is *fully complemented* if F is complemented in E and $F^\perp \approx E$.

COROLLARY 6.3. *If E is a countably generated Hilbert A-module then E is unitarily equivalent to a fully complemented submodule of H_A.*

This corollary (which is an immediate consequence of Theorem 6.2) provides an answer to the question posed at the beginning of the chapter.

It can also be stated in terms of projections. We define a projection in $\mathcal{L}(E)$ to be fully complemented if its range is a fully complemented subspace. Then the problem of classifying countably generated Hilbert A-modules is the same as the problem of classifying fully complemented projections in $\mathcal{L}(H_A)$ (up to unitary equivalence).

Using Theorem 6.2, we can clear up some unfinished business from Chapter 4, where we defined the exterior tensor product of Hilbert C*-modules but neglected to prove that the inner product on such a space is positive definite.

Let E be a Hilbert A-module, let F be a Hilbert B-module and suppose that

$$z = \sum_{i=1}^{k} x_i \otimes y_i \in E \otimes_{\mathrm{alg}} F.$$

We wish to show that if $\langle z, z\rangle = 0$ in $A \otimes_{\mathrm{alg}} B$ then $z = 0$ in $E \otimes_{\mathrm{alg}} F$. Note first that the condition $z = 0$ is independent of the ambient spaces E, F. That is, if E_1 is a sub-A-module of E and F_1 is a sub-B-module of F such that $z \in E_1 \otimes_{\mathrm{alg}} F_1$ then $z = 0$ as an element of $E_1 \otimes_{\mathrm{alg}} F_1$ if and only if $z = 0$ as an element of $E \otimes_{\mathrm{alg}} F$. The underlying algebraic reason for this is that short exact sequences of vector spaces always split; see p.161 of [Pie] for a clear account of this.

Replacing E by the submodule generated by $x_1, \ldots, x_k$, and F by the submodule generated by $y_1, \ldots, y_k$, we may assume that E and F are finitely generated. But then by Corollary 6.3 E embeds as a submodule of H_A and F as a submodule of H_B. So by the reasoning in the previous paragraph, we may assume that $E = H_A$ and $F = H_B$. As noted in Chapter 1 (and Chapter 3), H_A is unitarily equivalent to a direct sum $\bigoplus_{n=1}^{\infty} A$ of copies of A, so we may regard elements of H_A as being sequences of elements of A, and similarly for H_B.

So suppose that, for $1 \leqslant i \leqslant k$,

$$x_i = (a_{in}) \in H_A, \quad y_i = (b_{in}) \in H_B.$$

Then the condition $\langle z, z\rangle = 0$ becomes

$$0 = \Big\langle \sum_i x_i \otimes y_i, \sum_j x_j \otimes y_j \Big\rangle = \sum_{i,j,n} a_{in}^* a_{jn} \otimes b_{in}^* b_{jn}$$

$$= \sum_n \Big(\sum_i a_{in} \otimes b_{in}\Big)^* \Big(\sum_j a_{jn} \otimes b_{jn}\Big).$$

This implies that $\sum_i a_{in} \otimes b_{in} = 0$ for all n, in other words $z = 0$ as required.

Having at last justified the exterior tensor product construction, we can use it to explain one reason for the importance of the Hilbert A-module H_A. We noted in Chapter 4 that H_A can be regarded as the exterior tensor product of the Hilbert $\mathbb{C}$-module H and the Hilbert A-module A. As proved in Chapter 4, this implies that

$$\mathcal{K}(H_A) \cong \mathcal{K}(H) \otimes \mathcal{K}(A) \cong K \otimes A,$$

where $K = \mathcal{K}(H)$ is the C*-algebra of compact operators. Thus $\mathcal{K}(H_A)$ is what is called the stable algebra of A ([Bla], 5.1.1); and by Theorem 2.4 $\mathcal{L}(H_A) \cong M(K \otimes A)$, the stable multiplier algebra of A ([Bla], Definition 12.1.3). Thus H_A is intimately associated with two C*-algebras which play central roles in the K-theory of A.

Our next result also makes use of the exterior tensor product, and will be needed for the proof of Theorem 6.5. It says that given a C*-algebra B we can exhibit any separable C*-algebra as a C*-subalgebra of $\mathcal{L}(H_B)$.

LEMMA 6.4. *If A, B are C*-algebras and A is separable then there is a faithful nondegenerate $*$-homomorphism $\pi: A \to \mathcal{L}(H_B)$.*

Proof. Since A is separable, it has a faithful nondegenerate $*$-representation π_0 on the separable Hilbert space H, by [Ped], Corollary 3.7.5. For a in A, define $\pi(a) = \pi_0(a) \otimes 1$, where 1 is the identity element of $M(B)$. Then $\pi(a) \in \mathcal{L}(H) \otimes M(B)$ which, as in Chapter 4, we regard as a C*-subalgebra of $\mathcal{L}(H \otimes B)$. Thus π is a $*$-homomorphism from A to $\mathcal{L}(H_B)$, and it is easy to see that π is faithful and nondegenerate.

We are now ready for the next theorem of Kasparov. This says that under appropriate countability restrictions on the C*-algebras A and B, if $\rho: A \to \mathcal{L}(H_B)$ is nondegenerate and completely positive then A can be nondegenerately embedded in the 2×2 matrix algebra over $\mathcal{L}(H_B)$ in such a way that, for each a in A, $\rho(a)$ is the $(1,1)$-entry of the matrix representing the element a.

THEOREM 6.5. *Let A, B be C^*-algebras, with A separable and B σ-unital, and let $\rho: A \to \mathcal{L}(H_B)$ be a nondegenerate, completely positive mapping. Then there is a faithful, nondegenerate $*$-homomorphism π from A to the 2×2 matrix algebra $M_2(\mathcal{L}(H_B))$ such that $\rho(a) = (\pi(a))_{11}$ for all a in A.*

Proof. Let $(F_\rho, \pi_\rho, v_\rho)$ be the KSGNS construction associated with ρ, and note that v_ρ is an isometry since ρ is nondegenerate. Let $G = \operatorname{ran}(v_\rho)$ and let $p \in \mathcal{L}(F_\rho, G)$ be the projection from F_ρ onto G. Referring to the details of the KSGNS sconstruction, we see that $F_\rho \approx A \otimes_\rho H_B$, and since A is separable and H_B is countably generated it follows that F_ρ is also countably generated, and hence so is $G^\perp$. So by Theorem 6.2 we have $G^\perp \oplus H_B \approx H_B$. Let w be a unitary map from H_B onto $G^\perp \oplus H_B$. Then $pv_\rho \oplus w$ is a unitary map from $H_B \oplus H_B$ onto $G \oplus (G^\perp \oplus H_B)$, which we identify by associativity with $(G \oplus G^\perp) \oplus H_B = F_\rho \oplus H_B$. Let π_1 be a faithful, nondegenerate $*$-homomorphism from A to $\mathcal{L}(H_B)$ (which exists by Lemma 6.4), and for a in A let

$$\pi(a) = (v_\rho^* p^* \oplus w^*)(\pi_\rho(a) \oplus \pi_1(a))(pv_\rho \oplus w).$$

Then $\pi: A \to \mathcal{L}(H_B \oplus H_B)$ is a $*$-homomorphism which is faithful and nondegenerate since it is unitarily equivalent to $\pi_\rho \oplus \pi_1$. For x in H_B, it is trivially verified that

$$\pi(a)(x \oplus 0) = v_\rho^* \pi_\rho(a) v_\rho(x) \oplus z = \rho(a)x \oplus z,$$

where $z = w^*(1-p)\pi_\rho(a)v_\rho(x) \in H_B$. In terms of the representation of $\pi(a)$ as a 2×2 matrix over $\mathcal{L}(H_B)$, this tells us that $(\pi(a))_{11} = \rho(a)$.

The following schematic diagram may help to clarify the above construction.

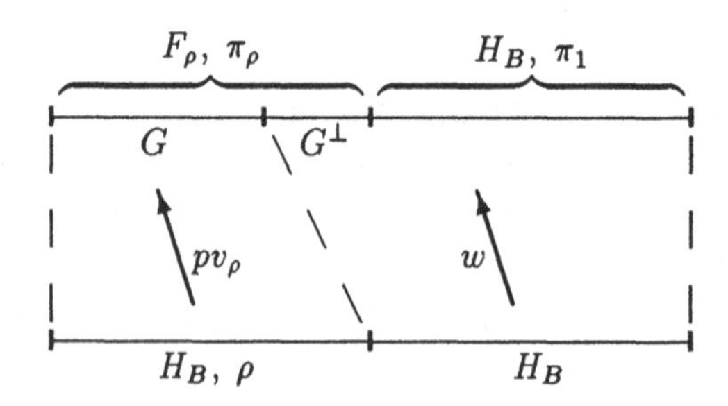

In the diagram, the lower line is intended to represent the direct sum of two copies of H_B, with $\rho(A)$ acting on the first. The upper line represents $F_\rho \oplus H_B$, with $\pi_\rho(A)$ acting on F_ρ and $\pi_1(A)$ acting on H_B.

If we do not assume that ρ is nondegenerate then we can still obtain a result like Theorem 6.5, but at the expense of having to drop the condition that π should be nondegenerate. We state this as a separate proposition.

PROPOSITION 6.6. *Let A, B be as in Theorem 6.5 and let $\rho: A \to \mathcal{L}(H_B)$ be a completely positive mapping. Then there is a faithful $*$-homomorphism $\pi: A \to M_2(\mathcal{L}(H_B))$ such that $\rho(a) = (\pi(a))_{11}$ for all a in A.*

Proof. As noted in Chapter 5, we can extend ρ to a completely positive unital map from A^+ to $\mathcal{L}(H_B)$, and it suffices to apply Theorem 6.5 to this extended map.

There is another way of looking at Theorem 6.5 (or Proposition 6.6), as follows. We write M_2 for $M_2(\mathbf{C})$, the algebra of 2×2 matrices. Since $H \oplus H$ is a separable Hilbert space, it is unitarily equivalent to H, and therefore

$$M_2 \otimes K \cong \mathcal{K}(H \oplus H) \cong \mathcal{K}(H) = K$$

(this elementary fact is at the heart of the whole of C*-algebraic K-theory and KK-theory). Let p denote the projection onto the first coordinate of $H_B \oplus H_B$, so that p is a fully complemented projection in $\mathcal{L}(H_B \oplus H_B)$. Under the isomorphisms

$$\begin{aligned}\mathcal{L}(H_B \oplus H_B) &\cong M(\mathcal{K}(H_B \oplus H_B)) \\ &\cong M(M_2(\mathcal{K}(H_B))) \\ &\cong M(M_2(K \otimes B)) \\ &\cong M(M_2 \otimes K \otimes B) \cong M(K \otimes B),\end{aligned}$$

we can regard p as a fully complemented projection in the stable multiplier algebra of B. In the same way, we can regard both ρ and π in Theorem 6.5 as mappings into $M(K \otimes B)$, and the theorem then says that ρ is unitarily equivalent to the compression of π to the range of p.

We conclude this chapter with a couple of miscellaneous results about σ-unital C*-algebras.

PROPOSITION 6.7. *A Hilbert A-module E is countably generated if and only if the C^*-algebra $\mathcal{K}(E)$ is σ-unital.*

Proof. As in the proof of Theorem 6.2, we may assume that A is unital and therefore H_A is countably generated, with a generating set consisting of the elements $e_n = \varepsilon_n \otimes 1$. The element $h = \sum_{n=1}^{\infty} 2^{-n} \theta_{e_n,e_n}$ of $\mathcal{K}(H_A)$ is strictly positive (for if ρ is a state of $\mathcal{K}(H_A)$ with $\rho(h) = 0$ then $\rho(\theta_{e_n,e_n}) = 0$ for all n; hence, by the Cauchy–Schwarz inequality, $\rho(\theta_{x,e_n}) = 0$ for all x in H_A, from which it is easy to see that $\rho(\mathcal{K}(H_A)) = \{0\}$). Hence $\mathcal{K}(H_A)$ is σ-unital.

Suppose that E is countably generated. By Corollary 6.3 we can identify E with a complemented submodule of H_A. Let $p \in \mathcal{L}(H_A, E)$ be the projection from H_A onto E. Given a countable approximate unit (u_n) for $\mathcal{K}(H_A)$, it is easy to see that (pu_np^*) is an approximate unit for $\mathcal{K}(E)$. So $\mathcal{K}(E)$ is σ-unital.

Conversely, suppose that $\mathcal{K}(E)$ is σ-unital, with countable approximate unit (v_n). For each n, we can find finite collections of elements $x(n,i)$, $y(n,i)$ $(1 \leqslant i \leqslant m_n)$ of E such that

$$\Big\| \sum_{i=1}^{m_n} \theta_{x(n,i),y(n,i)} - v_n \Big\| < \frac{1}{n}.$$

Given x in E and $\varepsilon > 0$, we have $v_n x \to x$ as $n \to \infty$ so, for n large enough,

$$\Big\| \sum_{i=1}^{m_n} \theta_{x(n,i),y(n,i)} x - x \Big\| < \varepsilon.$$

But $\theta_{x(n,i),y(n,i)} x = x(n,i)\langle y(n,i), x\rangle$. It is clear from this that the countable set $\{x(n,i) \colon 1 \leqslant i \leqslant m_n,\ n \geqslant 1\}$ generates E.

The result that follows is the noncommutative Tietze theorem first proved in [AkePedTom]. The proof that we give is not so very different from that in [Ped], Proposition 3.12.10. But there is some simplification, and this result seems to fit naturally into an account of Hilbert C*-modules.

We shall need to quote the following fact (Exercise 4.6.21 in [KadRin]). Let $\alpha \colon A \to B$ be a $*$-homomorphism between unital C*-algebras, and suppose that $0 \leqslant b \leqslant d$ in $\alpha(A)$, $0 \leqslant c \in A$ and $\alpha(c) = d$. Then there exists $a \in A$ with $0 \leqslant a \leqslant c$ such that $\alpha(a) = b$.

Proposition 6.8. *Let A, B be C^*-algebras with A σ-unital. If $\alpha: A \to B$ is a $*$-homomorphism which is surjective then the canonical extension $\bar{\alpha}: M(A) \to M(B)$ is also surjective.*

Proof. Note that since α is surjective it is nondegenerate, so that $\bar{\alpha}$ is defined (see Chapter 2) and is unital.

Let h be a strictly positive element of A and let (u_n) be a countable approximate unit for A. Let z be a positive element in the unit ball of $M(B)$. It will suffice to find x in $M(A)$ with $\bar{\alpha}(x) = z$.

Since $\alpha(u_n) \to 1$ strictly, it follows that $b_n \to z$ strictly, where $b_n \in B$ is given by $b_n = z^{\frac{1}{2}}\alpha(u_n)z^{\frac{1}{2}}$. In particular, the sequence $\big(b_n\alpha(h)\big)$ is Cauchy in B, and so for $m > n$

$$\|(b_m - b_n)^{\frac{1}{2}}\alpha(h)\|^2 = \|\langle\alpha(h), (b_m - b_n)\alpha(h)\rangle\| \to 0$$

as $m, n \to \infty$. Replacing (b_n) by a subsequence, we may assume that

$$\|(b_{n+1} - b_n)^{\frac{1}{2}}\alpha(h)\| < 2^{-n}.$$

We construct inductively a sequence (a_n) of positive elements of $M(A)$ satisfying

$$a_n \leqslant a_{n+1} \leqslant 1, \qquad \alpha(a_n) = b_n, \qquad \|(a_{n+1} - a_n)h\| < 2^{-n}. \tag{6.2}$$

If we can do this, then (a_nh) will be Cauchy in A. Since (a_n) is bounded and hA is dense in A (by Lemma 6.1), it will follow that (a_nc) is Cauchy, for all c in A. In other words, (a_n) is strictly Cauchy, and therefore has a strict limit x in $M(A)$. By strict continuity of $\bar{\alpha}$ we must have $\bar{\alpha}(x) = z$, as required.

To start the induction, choose any positive a_1 in the unit ball of A with $\alpha(a_1) = b_1$. Given a_n, choose c in $M(A)$ such that $0 \leqslant c \leqslant 1 - a_n$ and $\bar{\alpha}(c) = b_{n+1} - b_n$. (This is possible by the fact quoted just before the statement of the proposition, since $0 \leqslant b_{n+1} - b_n \leqslant 1 - b_n$ in the range of $\bar{\alpha}$.) We want to adjust c (by adding an element of $\ker(\bar{\alpha})$) in order to make $\|ch\|$ small. To do this, let (e_i) be an approximate unit (not necessarily countable) for $\ker(\bar{\alpha})$. By [KadRin] Exercise 4.6.60, or [Ped] Lemma 1.5.4,

$$\lim_i \|(1 - e_i)c^{\frac{1}{2}}h\| = \|\bar{\alpha}(c)^{\frac{1}{2}}\alpha(h)\| = \|(b_{n+1} - b_n)^{\frac{1}{2}}\alpha(h)\| < 2^{-n}.$$

Choose i so that $\|(1-e_i)c^{\frac{1}{2}}h\| < 2^{-n}$. Then $\|c^{\frac{1}{2}}(1-e_i)c^{\frac{1}{2}}h\| < 2^{-n}$ (since $\|c^{\frac{1}{2}}\| \leqslant 1$). Set $a_{n+1} = a_n + c^{\frac{1}{2}}(1-e_i)c^{\frac{1}{2}}$. Then $a_{n+1} \leqslant a_n + c \leqslant 1$; and all the other conditions in (6.2) are evidently satisfied also. That completes the inductive construction and the proof.

As noted in [Ped], 3.12.11, Proposition 6.8 fails (even in the commutative case) if we drop the requirement that A should be σ-unital. A more elaborate version of Proposition 6.8, due to Olsen and Pedersen, is proved in [JenTho], Theorem 1.1.26.

The material covered in this chapter has brought us perilously close to a study of K-theory and KK-theory. These topics are among the most important applications of the theory of Hilbert C*-modules, and indeed it was the application to KK-theory that motivated Kasparov to prove the fundamental theorems about Hilbert C*-modules. But it is not the aim of this book to go further in this direction, since there is already an excellent exposition of KK-theory in the book by Jensen and Thomsen [JenTho]. Readers who want to learn KK-theory will find that they need to generalise some results about Hilbert C*-modules to the context of graded C*-algebras, but that will not be hard for those who have understood the first six chapters of this book. They are strongly advised to read the relevant sections of [Bla] for an overview of what KK-theory is about, before plunging into the technicalities of [JenTho].

References for Chapter 6: [Kas 1], [MinPhi].

Chapter 7

Full modules, Morita equivalence

Many properties of a C*-algebra (anything K-theoretic, for instance) are unaffected if the algebra is tensored with K. Given a C*-algebra A, we call $K \otimes A$ the ***stable algebra*** of A. A C*-algebra B is called ***stable*** if $B \cong K \otimes B$. Of course, the stable algebra of A is always stable since $K \cong K \otimes K$. Two C*-algebras A, B are said to be ***stably isomorphic*** if their stable algebras are isomorphic. In this short chapter we shall prove a theorem of Brown, Green and Rieffel which characterises stable isomorphism for σ-unital C*-algebras in terms of a relation called Morita equivalence. We shall not investigate Morita equivalence in any systematic way, but just develop enough of its properties to prove the main theorem. For a fuller treatment, see [Rie 1], [Rie 2], [Bro], [BroGreRie].

Before defining Morita equivalence, we need to investigate some properties of full Hilbert C*-modules. A Hilbert A-module E is said to be ***full*** if the ideal $\langle E, E \rangle$ is dense in A.

Our first result about full Hilbert C*-modules involves a construction that will be relevant to the notion of Morita equivalence. Let A be a C*-algebra and let E, F be Hilbert A-modules. Under the natural operation of composition of mappings, $\mathcal{K}(E, F)$ is a right $\mathcal{K}(E)$-module. (In the same way, $\mathcal{K}(E, F)$ is a left $\mathcal{K}(F)$-module. This $\mathcal{K}(E)$-$\mathcal{K}(F)$-bimodule structure is central to the study of Morita equivalence.) Set $B = \mathcal{K}(E)$ and $G = \mathcal{K}(E, F)$. Then not only is G a right B-module but also it has a B-valued inner product, defined by

$$\langle s, t \rangle_B = s^* t \qquad (s, t \in G).$$

The properties (1.1) for an inner product are easily verified, as is the fact that for s in G the inner-product norm $\|\langle s,s\rangle_B\|^{\frac{1}{2}}$ is the same as the operator norm of s as an element of $\mathcal{K}(E,F)$. Thus G is complete and is therefore a Hilbert B-module.

PROPOSITION 7.1. *Let A be a C*-algebra, let E, F be Hilbert A-modules, let $B=\mathcal{K}_A(E)$ and let G be the Hilbert B-module $\mathcal{K}_A(E,F)$. If E is full then $\mathcal{K}_B(G)\cong\mathcal{K}_A(F)$ and $\mathcal{L}_B(G)\cong\mathcal{L}_A(F)$.*

Proof. For t in $\mathcal{L}_A(F)$ and u, v in G, we have

$$\langle tu,v\rangle_B=(tu)^*v=u^*t^*v=\langle u,t^*v\rangle_B.$$

Thus the map $\alpha(t)\colon u\mapsto tu\ \ (u\in G)$ is adjointable and therefore in $\mathcal{L}_B(G)$. In other words, the left $\mathcal{L}_A(F)$-module structure on G provides a map $\alpha\colon\mathcal{L}_A(F)\to\mathcal{L}_B(G)$, which is easily seen to be a $*$-homomorphism.

If $\alpha(t)=0$ then $tu=0$ for all u in G. In particular,

$$t\theta_{z,x}(y)=0\qquad(x,y\in E,\ z\in F),$$

so $tz\langle x,y\rangle=0$. Now suppose that E is full, so that

$$\overline{F\langle E,E\rangle}=\overline{FA}=F.$$

Since $tF\langle E,E\rangle=\{0\}$, this means that $t=0$. Thus α is injective.

Let $x,y\in E$, $z,w\in F$, and set $s=\theta_{z,x}$, $t=\theta_{w,y}$. Then $s,t\in G$, and by a routine calculation we have

$$\alpha\big(\theta_{z\langle x,y\rangle,w}\big)=\theta_{s,t}.\tag{7.1}$$

Since G is generated as a normed linear space by elements of the form s and t, and $*$-homomorphisms between C*-algebras have closed range, it follows from (7.1) that $\alpha\big(\mathcal{K}_A(F)\big)\supseteq\mathcal{K}_B(G)$. On the other hand, if E is full then elements of the form $\theta_{z\langle x,y\rangle,w}$ generate $\mathcal{K}_A(F)$, so (7.1) shows that $\alpha\big(\mathcal{K}_A(F)\big)\subseteq\mathcal{K}_B(G)$. Thus the restriction of α to $\mathcal{K}_A(F)$ provides a $*$-isomorphism between $\mathcal{K}_A(F)$ and $\mathcal{K}_B(G)$, and we have proved the first part of the proposition. The second part follows from the observation that if two C*-algebras are isomorphic then so are their multiplier algebras. (Alternatively, it is not hard to check that α is strictly continuous on the unit ball of $\mathcal{L}_A(F)$. So α is the canonical extension to $\mathcal{L}_A(E)$ of its restriction to $\mathcal{K}_A(F)$ and is therefore an isomorphism from $\mathcal{L}_A(F)$ onto $\mathcal{L}_B(G)$.)

The enquiring reader will already have asked the question: Why did we define $B = \mathcal{K}(E)$ and $G = \mathcal{K}(E,F)$ rather than using $\mathcal{L}(E)$ and $\mathcal{L}(E,F)$? (Any reader who is insufficiently curious to have wondered about this should instead be asking: Why am I reading this book at all?) The answer to the (first) question is that we can indeed make $\mathcal{L}(E,F)$ into a Hilbert $\mathcal{L}(E)$-module. But, as is already apparent from the proof of Proposition 7.1, the $\mathcal{K}$ spaces are technically easier to handle than the $\mathcal{L}$ spaces; and it will be clear when we come to study Morita equivalence that they are the appropriate objects of study.

The proof of Proposition 7.1 is conceptually a little bit sophisticated, with mappings $\theta_{s,t}$ where the subscripts s, t are themselves of the form $\theta_{z,x}$; but its mathematical content is almost trivial. By contrast, the next two lemmas, which are taken from [Bro], are conceptually straightforward but require a fair amount of technical agility to prove.

LEMMA 7.2. *Let E be a Hilbert A-module and let*

$$S = \Big\{c \in A\colon \|c\| \leqslant 1 \text{ and } c = \sum_{i=1}^{k} \langle x_i, x_i\rangle, \text{ where } k \geqslant 1,\ x_1, \ldots, x_k \in E\Big\}.$$

Then

(i) *given $y_1, \ldots, y_n$ in E and $\varepsilon' > 0$, there exists $c \in S$ with*

$$\|(1-c)|y_i|\,\| < \varepsilon' \qquad (1 \leqslant i \leqslant n);$$

(ii) *if E is full, $a \in A$ and $\varepsilon > 0$ then there exists $c \in S$ with*

$$\|(1-c)a\| < \varepsilon.$$

Proof. (i): This argument is a variant of that used in the standard construction of an approximate unit in a C*-algebra ([Dix 2], Proposition 1.7.2). Recalling that we can regard E as an A^+-module, where A^+ is the unitisation of A, define

$$x_i = y_i\Big(\varepsilon'^2 + \sum_{j=1}^{n} |y_j|^2\Big)^{-\frac{1}{2}} \qquad (1 \leqslant i \leqslant n),$$

$$c = \sum_{i=1}^{n} \langle x_i, x_i \rangle.$$

By functional calculus for the element $\sum_j |y_j|^2$ of A, we have

$$\|c\| = \left\| \left(\varepsilon'^2 + \sum |y_j|^2\right)^{-\frac{1}{2}} \sum |y_j|^2 \left(\varepsilon'^2 + \sum |y_j|^2\right)^{-\frac{1}{2}} \right\| \leqslant 1,$$

and so $c \in S$. By a similar use of the functional calculus,

$$\left\|(1-c) \sum |y_j|^2 (1-c)\right\| = \left\| f\left(\sum |y_j|^2\right) \right\| \leqslant \tfrac{1}{4}\varepsilon'^2,$$

where f is the function $t \mapsto \varepsilon'^4 t(\varepsilon'^2 + t)^{-2}$ on $\mathrm{sp}\left(\sum |y_j|^2\right)$. Thus

$$\|(1-c)|y_j|^2(1-c)\| \leqslant \tfrac{1}{4}\varepsilon'^2 \qquad (1 \leqslant j \leqslant n),$$

from which it follows that $\|(1-c)|y_j|\| \leqslant \frac{1}{2}\varepsilon' < \varepsilon'$.

(ii): Since E is full, we can approximate a by a finite linear combination of elements $\langle y_i, z_i \rangle$, where $y_i, z_i \in E$. By polarisation, we may assume that $z_i = y_i$. So suppose that $\|a - a'\| < \varepsilon$, where a' is a finite linear combination of elements $\langle y_i, y_i \rangle$. Now apply (i) with a small enough value of ε' to find c in S with $\|(1-c)a\| < \varepsilon$, as required.

LEMMA 7.3. *If A is a σ-unital C^*-algebra and E is a full Hilbert A-module then there is a sequence (x_n) in E such that $\sum \langle x_n, x_n \rangle$ converges strictly to 1 in $M(A)$.*

Proof. It will be sufficient to find a sequence (c_n) in S with $\sum c_n$ converging strictly to 1.

Let h be a strictly positive element of A. We construct inductively a sequence (c_j) in S such that

$$\sum_{j=1}^{n} c_j \leqslant 1, \qquad \left\| \left(1 - \sum_{j=1}^{n} c_j\right) h \right\| < 2^{-n}. \tag{7.2}$$

To start the induction, use Lemma 7.2(ii) to find c_1 in S such that $\|(1 - c_1)h\| < \frac{1}{2}$. Supposing that $c_1, \ldots, c_n$ have been chosen, use Lemma 7.2(ii) to find d in S with

$$\left\|(1-d)\left(1 - \sum_{j=1}^{n} c_j\right)^{\frac{1}{2}} h\right\| < 2^{-n-1}.$$

Let

$$c_{n+1} = \Big(1 - \sum_{j=1}^{n} c_j\Big)^{\frac{1}{2}} d \Big(1 - \sum_{j=1}^{n} c_j\Big)^{\frac{1}{2}}.$$

Then $c_{n+1} \in S$ (for if $c = \sum\langle x_i, x_i\rangle \in S$ and $a \in A^+$ with $\|a\| \leqslant 1$ then $\|a^*ca\| \leqslant 1$ and $a^*ca = \sum\langle x_i a, x_i a\rangle$, so $a^*ca \in S$). Since $\|d\| \leqslant 1$, we have $c_{n+1} \leqslant 1 - \sum_{j=1}^{n} c_j$, so that $\sum_{j=1}^{n+1} c_j \leqslant 1$. Also,

$$\Big\|\Big(1 - \sum_{j=1}^{n+1} c_j\Big)h\Big\| = \Big\|\Big(1 - \sum_{j=1}^{n} c_j\Big)^{\frac{1}{2}}(1-d)\Big(1 - \sum_{j=1}^{n} c_j\Big)^{\frac{1}{2}} h\Big\| < 2^{-n-1}.$$

Thus (7.2) holds with $n+1$ in place of n and the inductive construction is complete.

Therefore $\sum_{n=1}^{\infty} c_n h = h$, so $\sum_{n=1}^{\infty} c_n h a = ha$ for all a in A. It follows from Lemma 6.1 that $\sum_{n=1}^{\infty} c_n a = a$ for all a in A, so that $\sum c_n$ converges strictly to 1 in $M(A)$, as required.

PROPOSITION 7.4. *Let A be a σ-unital C^*-algebra, let E be a full Hilbert A-module and let H be a separable Hilbert space. Then*

(i) *there is a Hilbert A-module F such that $H \otimes E \approx A \oplus F$;*
(ii) *if E is countably generated, $H \otimes E \approx H_A$.*

Proof. (i): Using Lemma 7.3, choose a sequence (x_n) in E such that $\sum\langle x_n, x_n\rangle = 1$ strictly in $M(A)$; and let (ε_n) be an orthonormal basis for H. Define $t: A \to H \otimes E$ by $t(a) = \sum_n \varepsilon_n \otimes x_n a$ $(a \in A)$. Note that

$$\langle t(a), t(a)\rangle = \sum_n a^*\langle x_n, x_n\rangle a = a^*a$$

and, since the sum converges, $t(a)$ is a well-defined element of $H \otimes E$. A simple computation shows that t is adjointable, with

$$t^*(\varepsilon_n \otimes y) = \langle x_n, y\rangle,$$

and that t^*t is the identity on A. Thus t is an isometry, and we can take $F = \ker(t^*)$ as the orthogonal complement for the copy $t(A)$ of A in $H \otimes E$.

(ii): The proof is a sequence of unitary equivalences, the last of which uses Kasparov's stabilisation theorem (along the way there occur a distributivity relation between direct sums and tensor products and also an associativity relation for tensor products, which the careful reader will of course justify):

$$H \otimes E \approx H \otimes H \otimes E \approx H \otimes (A \oplus F) \approx (H \otimes A) \oplus (H \otimes F)$$
$$= H_A \oplus (H \otimes F) \approx H_A.$$

The result of Proposition 7.4(ii) can be expressed as follows. If A is σ-unital and E is a Hilbert A-module which is "large enough" (full) but "not too large" (countably generated) then E is stably unitarily equivalent to A, where "stable" in this context means "when tensored with H".

We are now ready for Morita equivalence. We shall say that two C*-algebras A, B are *Morita equivalent*, written $A \sim_M B$, if there is a full Hilbert A-module E such that $B \cong \mathcal{K}_A(E)$. (In the literature, this relation is usually referred to as strong Morita equivalence.)

PROPOSITION 7.5. *Morita equivalence is an equivalence relation.*

Proof. The relation is reflexive since $A \cong \mathcal{K}_A(A)$. It is symmetric since, by Proposition 7.1 (with $F = A$), if $B \cong \mathcal{K}_A(E)$ and G is the Hilbert B-module $\mathcal{K}_A(E, A)$ (exercise: G is full if E is) then $A \cong \mathcal{K}_B(G)$.

To see that Morita equivalence is transitive, suppose that $B \cong \mathcal{K}_A(E)$ and $C \cong \mathcal{K}_B(F)$, where E is a full Hilbert A-module and F is a full Hilbert B-module. Let $\iota: B \to \mathcal{L}_A(E)$ be the $*$-homomorphism arising from the natural embedding of $\mathcal{K}_A(E)$ in $\mathcal{L}_A(E)$ and let $G = F \otimes_\iota E$. Then G is a full (check this!) Hilbert A-module, and $\iota_*: C \to \mathcal{K}_A(G)$ is a $*$-isomorphism by Proposition 4.7.

Now for the punch line:

THEOREM 7.6. *Two σ-unital C*-algebras are stably isomorphic if and only if they are Morita equivalent.*

Proof. Observe first that for any C*-algebra A,

$$\mathcal{K}_A(H_A) = \mathcal{K}_A(H \otimes A) \cong \mathcal{K}_{\mathbf{C}}(H) \otimes \mathcal{K}_A(A) = K \otimes A,$$

so that $A \sim_M K \otimes A$. If A and B are stably isomorphic then

$$A \sim_M K \otimes A \sim_M K \otimes B \sim_M B,$$

hence $A \sim_M B$ by Proposition 7.5. Note that this part of the proof holds even without the hypothesis that A, B are σ-unital.

For the converse, suppose that $A \sim_M B$, say $B \cong \mathcal{K}_A(E)$ where E is a full Hilbert A-module. If A and B are σ-unital then E must be countably generated (by Proposition 6.7), so by Proposition 7.4

$$K \otimes B \cong \mathcal{K}_A(H \otimes E) \cong \mathcal{K}_A(H_A) \cong K \otimes A.$$

References for Chapter 7: [Rie 1], [Bro], [BroGreRie], [MinPhi].

Chapter 8

Slice maps and bialgebras

An important technical tool throughout our treatment of Hilbert C*-modules has been the strict topology on $\mathcal{L}(E)$, where E as usual is a Hilbert A-module. There is an ambiguity in the definition of the strict topology, which we want to explain and then (partially) resolve. When we say that a directed net (t_i) in $\mathcal{L}(E)$ converges strictly to t, we mean that

$$t_i x \xrightarrow{i} tx, \quad t_i^* x \xrightarrow{i} t^* x \qquad (x \in E). \tag{8.1}$$

In the particular case when $E = A$, we obtain the definition of strict convergence in the multiplier algebra $M(A)$. But, for a general Hilbert A-module E, $\mathcal{L}(E)$ is the multiplier algebra of $\mathcal{K}(E)$. To say that $t_i \to t$ in the strict topology of $\mathcal{L}(E)$ considered as the multiplier algebra of $\mathcal{K}(E)$ means that

$$t_i k \xrightarrow{i} tk, \quad t_i^* k \xrightarrow{i} t^* k \qquad (k \in \mathcal{K}(E)), \tag{8.2}$$

where the convergence this time is in the $\mathcal{L}(E)$ norm. *In general, the topologies on $\mathcal{L}(E)$ given by* (8.1) *and* (8.2) *are different.* To see this, denote by σ_1, σ_2 the topologies on $\mathcal{L}(E)$ given by (8.1) and (8.2) respectively, and take $A = \mathbf{C}$ so that E is a Hilbert space H. Then σ_1 is by definition the strong* topology on $\mathcal{L}(H)$. It can be shown that σ_2 is the ultrastrong* topology. Rather than attempting to justify or even explain this assertion, which follows from the results in [Tay], we give an explicit construction (taken from [Neu]) of a directed net in $\mathcal{L}(H)$ which converges to 0 for the σ_1 topology but not for σ_2.

Let P be the set of finite-rank projections in $\mathcal{L}(H)$, ordered by inclusion of ranges, and for p in P let $\operatorname{tr} p$ be the trace of p (the dimension of the range of p). Let $t_p = (\operatorname{tr} p)(1-p)$. If $\xi_1, \ldots, \xi_n \in H$ and $\{\xi_1, \ldots, \xi_n\} \subseteq \operatorname{ran} p$ then $t_p \xi_i = 0 \quad (1 \leqslant i \leqslant n)$. Thus the directed net $(t_p)_{p \in P}$ converges to 0 in σ_1.

Let $\{\varepsilon_r\}_{r \geqslant 1}$ be an orthonormal basis for H, and define k in $\mathcal{K}(H)$ by $k\varepsilon_r = r^{-1}\varepsilon_r \quad (r \geqslant 1)$. Fix δ with $0 < \delta < 1$. For p in P we have

$$\sum_{r=1}^{\infty} \langle \varepsilon_r, p\varepsilon_r \rangle = \operatorname{tr} p,$$

and it follows that $\langle \varepsilon_{r_0}, p\varepsilon_{r_0} \rangle \leqslant \delta$ for some r_0 with $r_0 \leqslant \delta^{-1} \operatorname{tr} p$. Thus

$$\begin{aligned} \|t_p k\| \geqslant \|t_p k \varepsilon_{r_0}\| = r_0^{-1} \|t_p \varepsilon_{r_0}\| &= r_0^{-1} \operatorname{tr} p \, \|(1-p)\varepsilon_{r_0}\| \\ &= r_0^{-1} \operatorname{tr} p \, \langle \varepsilon_{r_0}, (1-p)\varepsilon_{r_0} \rangle^{\frac{1}{2}} \\ &\geqslant r_0^{-1} \operatorname{tr} p \, (1-\delta)^{\frac{1}{2}} \geqslant \delta(1-\delta)^{\frac{1}{2}}. \end{aligned}$$

Since this holds for all p in P, $(t_p k)$ does not converge to 0 in norm, and so (t_p) does not converge to 0 in σ_2.

Despite the ambiguity, we shall continue to refer to the strict topology on $\mathcal{L}(E)$ without specifying whether we mean σ_1 or σ_2. The reason for this cavalier attitude is that we are only ever interested in the strict topology on the unit ball of $\mathcal{L}(E)$, where the two topologies coincide. This fact is easily proved, but is of sufficient importance to be worth stating as a formal proposition:

PROPOSITION 8.1. *Let E be a Hilbert A-module. On the unit ball of $\mathcal{L}(E)$, the topologies given by* (8.1) *and* (8.2) *coincide.*

Proof. Let (t_i) be a directed net in the unit ball of $\mathcal{L}(E)$. If $t_i \to t$ in σ_1 then it is easy to see that $t_i\theta_{x,y} \to t\theta_{x,y}$ for all $x, y \in E$. By continuity and the boundedness of (t_i) it follows that $t_i k \to tk$ for all k in $\mathcal{K}(E)$. Repeating the argument with t_i^* in place of t_i, we see that $t_i \to t$ in σ_2.

Conversely, if $t_i \to t$ in σ_2 then $t_i\theta_{x,y}z \to t\theta_{x,y}z$ for all $x, y, z \in E$. Since $E\langle E, E \rangle$ is dense in E it follows as in the first part of the proof that $t_i \to t$ in σ_1.

We now turn to the first main theme of this chapter, which is the relation between completely positive mappings and tensor products. In Chapter 5 we considered strict completely positive mappings $\rho: A \to \mathcal{L}(F)$, where F

is a Hilbert B-module. Now we shall restrict attention to the case $F = B$, and look at mappings $\rho: A \to M(B)$. This apparent loss of generality is not real: for if F is a Hilbert B-module then $\mathcal{L}(F) \cong M(\mathcal{K}(F))$ and, in view of Proposition 8.1, $\rho: A \to \mathcal{L}(F)$ will be strict if and only if $\rho: A \to M(\mathcal{K}(F))$ is strict. So we can replace B by $\mathcal{K}(F)$ and regard ρ as a map from A to $M(B)$.

PROPOSITION 8.2. *Let A, A_0, B, B_0 be C*-algebras and let $\rho: A \to M(A_0)$, $\sigma: B \to M(B_0)$ be strict completely positive mappings. There is a unique strict completely positive map $\rho\otimes\sigma: A\otimes B \to M(A_0\otimes B_0)$ such that*

$$(\rho\otimes\sigma)(a\otimes b) = \rho(a)\otimes\sigma(b) \qquad (a \in A,\ b \in B).$$

If ρ, σ are nondegenerate (resp. retractions) then so is $\rho\otimes\sigma$.

Proof. Let $(F_\rho, \pi_\rho, v_\rho)$, $(F_\sigma, \pi_\sigma, v_\sigma)$ be the KSGNS constructions associated with ρ, σ. It follows from the results of Chapter 4 that we can form the nondegenerate $*$-homomorphism $\pi_\rho\otimes\pi_\sigma: A\otimes B \to \mathcal{L}_{A_0\otimes B_0}(F_\rho\otimes F_\sigma)$. For x in $A\otimes B$, define

$$(\rho\otimes\sigma)x = (v_\rho\otimes v_\sigma)^* \cdot (\pi_\rho\otimes\pi_\sigma)x \cdot (v_\rho\otimes v_\sigma).$$

By Proposition 5.5, $\rho\otimes\sigma$ has the required properties. The uniqueness of $\rho\otimes\sigma$ is obvious, and the assertions about nondegeneracy and retractions are easily verified.

In the above proof, we have $v_\rho \in \mathcal{L}_{A_0}(A_0, F_\rho)$, $v_\sigma \in \mathcal{L}_{B_0}(B_0, F_\sigma)$ and $v_\rho\otimes v_\sigma \in \mathcal{L}_{A_0\otimes B_0}(A_0\otimes B_0, F_\rho\otimes F_\sigma)$. The vigilant reader may object that in Chapter 4 we defined the tensor product $s\otimes t$ in the case when each of s, t is an adjointable operator on a single Hilbert C*-module but not in the case of operators like v_ρ, v_σ which act between different modules. In fact, the construction in Chapter 4 works just as well in this more general setting, but only the most masochistic reader will wish to go back and check this in detail.

Proposition 8.2 still holds if we remove the word "strict" from both the hypothesis and the conclusion. In this case, we just use the non-strict version of the KSGNS construction described towards the end of Chapter 5.

Specialising the construction in Proposition 8.2 to the case where $A_0 = A$, ρ is the identity mapping ι on A, and $B_0 = \mathbb{C}$, we obtain, for each state σ of B, a retraction $\iota\otimes\sigma$ from $A\otimes B$ to $M(A)$ (identified with the subalgebra $M(A)\otimes 1$ of $M(A\otimes B)$). This retraction is called the right slice map associated with σ (compare [Tak], Corollary 4.2.5). In fact, since $(\iota\otimes\sigma)(a\otimes b) = \sigma(b)a \quad (a \in A,\ b \in B)$, the range of $\iota\otimes\sigma$ is contained in A. Similarly, for each state ρ of A we have a left slice map $\rho\otimes\iota: A\otimes B \to B$.

Since each element ω of the space B^* of bounded linear functionals on B is a linear combination of four states, we can use the appropriate linear combination of slice maps to define a bounded (in fact, completely bounded) linear map $\iota\otimes\omega: A\otimes B \to A$, having an extension $\iota\otimes\omega: M(A\otimes B) \to M(A)$ which is strictly continuous on the unit ball. For each χ in A^*, we have

$$\chi\big((\iota\otimes\omega)x\big) = (\chi\otimes\omega)x \qquad (x \in A\otimes_{\rm alg}B).$$

Hence $\big|\chi\big((\iota\otimes\omega)x\big)\big| \leqslant \|\chi\|\,\|\omega\|\,\|x\|$ since $\|\chi\otimes\omega\| = \|\chi\|\,\|\omega\|$. It follows that $\|\iota\otimes\omega\| = \|\omega\|$.

The above construction will most frequently be used when A, B are C*-algebras of the form $\mathcal{K}(E)$, for some Hilbert C*-module E. In that context, it is convenient to use the following notation. We denote by $\mathcal{L}(E)_*$ the set of all bounded linear functionals on $\mathcal{L}(E)$ that are strictly continuous on the unit ball. It is easy to see that this is the same as the set of all strictly continuous extensions to $\mathcal{L}(E)$ of elements of $\mathcal{K}(E)^*$. Then we can state what we have just proved in the form of the following proposition.

PROPOSITION 8.3. *Let E, F be Hilbert C*-modules and let $\omega \in \mathcal{L}(F)_*$. There is a bounded linear map $\iota\otimes\omega: \mathcal{L}(E\otimes F) \to \mathcal{L}(F)$ with $\|\iota\otimes\omega\| = \|\omega\|$ which is strictly continuous on the unit ball and satisfies*

$$(\iota\otimes\omega)(s\otimes t) = \omega(t)s \qquad (s \in \mathcal{L}(E),\ t \in \mathcal{L}(F)).$$

We call the map $\iota\otimes\omega$ a *right slice map*. Naturally, there is an analogous result for left slice maps $\chi\otimes\iota$, where $\chi \in \mathcal{L}(E)_*$. We can also define slice maps on exterior tensor products of three or more spaces. For example, if E, F, G are Hilbert C*-modules, $\rho \in \mathcal{L}(E)_*$, $\sigma \in \mathcal{L}(F)_*$ and $\tau \in \mathcal{L}(G)_*$ then we can define maps such as

$$\iota\otimes\sigma\otimes\iota : \mathcal{L}(E\otimes F\otimes G) \to \mathcal{L}(E\otimes G),$$
$$\rho\otimes\iota\otimes\tau : \mathcal{L}(E\otimes F\otimes G) \to \mathcal{L}(F),$$

and so on. These maps will be strictly continuous on the unit ball, and will take the obvious values on simple tensors:

$$(\iota\otimes\sigma\otimes\iota)(r\otimes s\otimes t) = \sigma(s) r\otimes t, \quad (\rho\otimes\iota\otimes\tau)(r\otimes s\otimes t) = \rho(r)\tau(t)s.$$

For $1 \leqslant i \leqslant n$, let E_i be a Hilbert A_i-module, and let F be the Hilbert $(A_1\otimes\cdots\otimes A_n)$-module $E_1\otimes\cdots\otimes E_n$. For each permutation σ in S_n, there is a natural unitary mapping u_σ from F onto the Hilbert $(A_{\sigma(1)}\otimes\cdots\otimes A_{\sigma(n)})$-module $E_{\sigma(1)}\otimes\cdots\otimes E_{\sigma(n)}$ given on simple tensors by

$$u_\sigma(x_1\otimes\cdots\otimes x_n) = x_{\sigma(1)}\otimes\cdots\otimes x_{\sigma(n)} \qquad (x_i \in E_i,\ 1 \leqslant i \leqslant n).$$

Given t in $\mathcal{L}(E_1\otimes E_2)$, define t_{12} in $\mathcal{L}(F)$ by

$$t_{12} = t\otimes 1 \in \mathcal{L}\big((E_1\otimes E_2)\otimes(E_3\otimes\cdots\otimes E_n)\big).$$

Given distinct indices j, k in $\{1,\ldots,n\}$ and t in $\mathcal{L}(E_j\otimes E_k)$, choose $\sigma \in S_n$ with $\sigma(1) = j$ and $\sigma(2) = k$ and define t_{jk} in $\mathcal{L}(F)$ by

$$t_{jk} = u_\sigma t_{12} u_\sigma^{-1}.$$

Informally, t_{jk} is the operator that acts like t on the j and k component spaces of F and like the identity on the other components (and it depends only on t, j and k, not on the particular permutation σ). In particular, if t is the simple tensor $t_j\otimes t_k$ (where $t_j \in \mathcal{L}(E_j)$, $t_k \in \mathcal{L}(E_k)$) then t_{jk} is the simple tensor $t_1\otimes\cdots\otimes t_n$, where $t_i = 1 \ \ (i \notin \{j,k\})$.

This so-called leg notation for describing operators on multiple tensor product spaces is very useful and versatile. It extends easily to the following slightly more general situation. If $1 \leqslant r \leqslant n$ and $t \in \mathcal{L}(E_1\otimes\cdots\otimes E_r)$ then let

$$t_{1\cdots r} = t\otimes 1 \in \mathcal{L}(F) \cong \mathcal{L}\big((E_1\otimes\cdots\otimes E_r)\otimes(E_{r+1}\otimes\cdots\otimes E_n)\big);$$

and if $\{k_1,\ldots,k_r\}$ is an r-tuple of distinct elements of $\{1,\ldots,n\}$ and $t \in \mathcal{L}(E_{k_1}\otimes\cdots\otimes E_{k_r})$ then let $t_{k_1\cdots k_r} = u_\sigma t_{1\cdots r} u_\sigma^{-1}$ where $\sigma \in S_n$ and $\sigma(i) = k_i \ \ (1 \leqslant i \leqslant r)$.

PROPOSITION 8.4. *Let E, F, G be Hilbert C^*-modules with $s \in \mathcal{L}(E\otimes F)$, $t \in \mathcal{L}(E\otimes G)$, $\rho \in \mathcal{L}(F)_*$, $\sigma \in \mathcal{L}(G)_*$. Then*

$$(\iota\otimes\rho\otimes\sigma)(s_{12}t_{13}) = (\iota\otimes\rho)s \cdot (\iota\otimes\sigma)t. \tag{8.3}$$

Proof. If s and t are simple tensors, say $s = s_1 \otimes s_2$, $t = t_1 \otimes t_3$, then $s_{12}t_{13} = (s_1t_1) \otimes s_2 \otimes t_3$, and

$$(\iota \otimes \rho \otimes \sigma)(s_{12}t_{13}) = \rho(s_2)s_1\sigma(t_3)t_1 = (\iota \otimes \rho)s \cdot (\iota \otimes \sigma)t,$$

as required. By taking linear combinations of simple tensors and using continuity, one sees that (8.3) holds whenever $s \in \mathcal{K}(E \otimes F)$ and $t \in \mathcal{K}(E \otimes G)$. But both sides of (8.3) are strictly continuous on the unit ball as functions of s and t, by Proposition 8.3. It follows from Proposition 1.3 that (8.3) holds for all s in $\mathcal{L}(E \otimes F)$ and t in $\mathcal{L}(E \otimes G)$.

Let A, B be C*-algebras, let $\alpha \in \mathrm{Mor}(A, B)$ and let H be a Hilbert space. Recall from Chapter 4 that we identify $(H \otimes A) \otimes_\alpha B$ with $H \otimes B$ via the unitary mapping

$$u\colon (\xi \otimes a) \dot{\otimes} b \mapsto \xi \otimes \alpha(a)b \qquad (\xi \in H,\ a \in A,\ b \in B).$$

The induced map $\alpha_*\colon \mathcal{L}(H \otimes A) \to \mathcal{L}(H \otimes B)$ is computed on simple tensors as follows. Let $t \in \mathcal{L}(H)$, $x \in M(A)$. Then

$$\begin{aligned}
\alpha_*(t \otimes x)(\xi \otimes \alpha(a)b) &= u\big([(t \otimes x) \cdot (\xi \otimes a)] \dot{\otimes} b\big) \\
&= u\big((t\xi \otimes xa) \dot{\otimes} b\big) \\
&= t\xi \otimes \alpha(xa)b \\
&= \big(t \otimes \alpha(x)\big) \cdot \big(\xi \otimes \alpha(a)b\big).
\end{aligned}$$

Thus $\alpha_*(t \otimes x) = t \otimes \alpha(x)$. Now let $\omega \in B^*$. Then

$$(\iota \otimes \omega)\alpha_*(t \otimes x) = (\iota \otimes \omega)(t \otimes \alpha(x)) = \omega(\alpha(x))t = (\iota \otimes \omega\alpha)(t \otimes x)$$

(where $\omega\alpha$ denotes the composition of ω with α). An argument like the proof of Proposition 8.4, using strict continuity, shows that

$$(\iota \otimes \omega)\alpha_* = \iota \otimes \omega\alpha \tag{8.4}$$

as maps from $\mathcal{L}(H \otimes A)$ to $\mathcal{L}(H)$.

With these results about slice maps, we are now ready to define C*-bialgebras, and thereby to introduce the topic that will dominate the remainder of the book. That is the study of C*-algebras associated with locally compact groups, and the more general algebras that go under the trendy name of C*-algebraic quantum groups. We shall barely scratch the surface of this subject, but we want to indicate some of the ways in which Hilbert C*-modules are likely to figure in any future formulation of an axiomatic basis for C*-algebraic quantum group theory.

A *comultiplication* on a C*-algebra A is an element δ in $\mathrm{Mor}(A, A\otimes A)$ such that

(i) $(\iota\otimes\delta)\delta = (\delta\otimes\iota)\delta$,

(ii) for all a, b in A, the elements $\delta(a)\cdot b\otimes 1$ and $\delta(a)\cdot 1\otimes b$ are in $A\otimes A$.

A C*-algebra with a comultiplication is called a *C*-bialgebra*. If (A_1, δ_1) and (A_2, δ_2) are C*-bialgebras then a *morphism* from (A_1, δ_1) to(A_2, δ_2) is a morphism $\alpha \in \mathrm{Mor}(A_1, A_2)$ such that $(\alpha\otimes\alpha)\delta_1 = \delta_2\alpha$.

Some comments on the above definition are in order. In condition (i), the products $(\iota\otimes\delta)\delta$, $(\delta\otimes\iota)\delta$ are products of morphisms as defined in Chapter 2. So if A is nonunital one must use the extensions of $\iota\otimes\delta$ and $\delta\otimes\iota$ to maps from $M(A\otimes A)$ to $M(A\otimes A\otimes A)$ in order to form their compositions with δ. If A is unital then δ maps A into $A\otimes A$ and condition (i) says that the diagram

$$\begin{array}{ccc} A & \xrightarrow{\delta} & A\otimes A \\ \downarrow{\scriptstyle\delta} & & \downarrow{\scriptstyle\delta\otimes\iota} \\ A\otimes A & \xrightarrow{\iota\otimes\delta} & A\otimes A\otimes A \end{array}$$

commutes. This is just the usual algebraic definition of coassociativity for a comultiplication. Condition (ii) (which is vacuous if A is unital) says that $\delta(a)$ is not just a multiplier of $A\otimes A$ but a bit more: it multiplies the subalgebras $1\otimes A$, $A\otimes 1$ of $M(A\otimes A)$ into $A\otimes A$.

Our first result about C*-bialgebras is that a comultiplication on a C*-algebra makes the dual space into an algebra. After proving this, we shall look in some detail at the two classical examples of C*-bialgebras, namely the function and (reduced) convolution C*-algebras of a locally compact group.

PROPOSITION 8.5. *If (A, δ) is a C*-bialgebra then the dual space A^* is an associative algebra under the product given by*

$$\phi \times \psi = \phi\otimes\psi \cdot \delta \qquad (\phi, \psi \in A^*).$$

Proof. For θ, ϕ, ψ in A^*, we have

$$(\theta \times \phi) \times \psi = \big(\big((\theta \otimes \phi) \cdot \delta\big) \otimes \psi\big) \cdot \delta = (\theta \otimes \phi \otimes \psi) \cdot (\delta \otimes \iota)\delta.$$

In the same way, $\theta \times (\phi \times \psi) = (\theta \otimes \phi \otimes \psi) \cdot (\iota \otimes \delta)\delta$. So the result follows from (i).

For a locally compact group G, let $A = C_0(G)$. We call A the *function C*-algebra of G*. We identify $A \otimes A$ with $C_0(G \times G)$ and $M(A \otimes A)$ with $C_b(G \times G)$. For f in A we define δf in $C_0(G \times G)$ by

$$\delta f(s,t) = f(st) \qquad (s,t \in G). \tag{8.5}$$

That δ satisfies (i) is an easy consequence of the associativity of the multiplication on G. For (ii), it is also easily verified that if $f, g \in A$ then the functions

$$(s,t) \mapsto f(st)g(t), \quad (s,t) \mapsto f(st)g(s)$$

are in $C_0(G \times G)$. Thus (A, δ) is a C*-bialgebra. Note that if G is noncompact and f is nonzero then δf is not in $C_0(G \times G)$ (because it is constant on "diagonal lines" $st = k$), but multiplication by $1 \otimes g$ or $g \otimes 1$ brings it into $C_0(G \times G)$. This "residual vanishing at infinity" (a description due to Iain Raeburn) is what is encapsulated by condition (ii).

The second classical example of a C*-bialgebra is the reduced convolution algebra of a locally compact group G. We refer to the Appendix to this chapter for basic properties of the C*-algebra $A = C_r^*(G)$, and we shall use the notation established there. In particular, dt will denote a right (not left) Haar measure on G. We identify $L^2(G \times G)$ with $L^2(G) \otimes L^2(G)$ in the usual way. Note that if $p, q, s, t \in G$ and $\xi \in L^2(G \times G)$ then $(\rho_p \otimes \rho_q \cdot \xi)(s,t) = \xi(sp, tq)$ (this is easily verified for simple tensors $\xi = \eta \otimes \zeta$, and is therefore true for general ξ by linearity and continuity).

Define a unitary operator w on $L^2(G \times G)$ by

$$w\xi(s,t) = \xi(st, t) \qquad (\xi \in L^2(G),\ s,t \in G),$$

and define a map $\delta_\rho : \mathcal{L}\big(L^2(G)\big) \to \mathcal{L}\big(L^2(G \times G)\big)$ by

$$\delta_\rho(x) = w^*(1 \otimes x)w \qquad \big(x \in \mathcal{L}\big(L^2(G)\big)\big).$$

Then, for r, s, t in G and ξ in $L^2(G\times G)$,

$$\begin{aligned}\delta_\rho(\rho_r)\xi(s,t) &= w^*(1\otimes\rho_r)w\xi(s,t)\\ &= (1\otimes\rho_r)w\xi(st^{-1},t)\\ &= w\xi(st^{-1},tr)\\ &= \xi(sr,tr) = (\rho_r\otimes\rho_r)\xi(s,t).\end{aligned}$$

Hence $\delta_\rho(\rho_r)=\rho_r\otimes\rho_r$.

For h in $C_{00}(G\times G)$ we define an operator t_h in $\mathcal{L}(L^2(G\times G))$ by

$$t_h = \iint h(s,t)\rho_s\otimes\rho_t\,ds\,dt. \tag{8.6}$$

Just as in the case of operators on $L^2(G)$ (see (A11) in the Appendix), this vector integral is to be interpreted in the sense of the weak operator topology, so (8.6) is an abbreviation for

$$\langle\xi,t_h\eta\rangle = \iint h(s,t)\langle\xi,\,\rho_s\otimes\rho_t\cdot\eta\rangle\,ds\,dt \qquad (\xi,\eta\in L^2(G\times G)).$$

It is easy to check that $\|t_h\|\leqslant\|h\|_1=\iint|h(s,t)|\,ds\,dt$. If

$$h=\sum_{i=1}^n f_i\otimes g_i\in C_{00}(G)\otimes_{\mathrm{alg}}C_{00}(G)$$

then $t_h=\sum_i\rho(f_i)\otimes\rho(g_i)\in\rho(A)\otimes\rho(A)$. Since every h in $C_{00}(G\times G)$ can be approximated in the $L^1(G\times G)$ norm by functions of this kind (Stone–Weierstrass theorem), it follows that $t_h\in\rho(A)\otimes\rho(A)$ for all h in $C_{00}(G\times G)$. Since on the other hand $\{\rho(f):f\in C_{00}(G)\}$ is dense in $\rho(A)$, it also follows that the set $\{t_h:h\in C_{00}(G\times G)\}$ is dense in $\rho(A)\otimes\rho(A)$.

For f in $C_{00}(G)$ and h in $C_{00}(G\times G)$, we have (using (A2))

$$\begin{aligned}\delta_\rho\big(\rho(f)\big)\cdot t_h &= \iiint f(r)h(s,t)\rho_{rs}\otimes\rho_{rt}\,dr\,ds\,dt\\ &= \iiint\Delta(r)^2f(r)h(r^{-1}s,r^{-1}t)\rho_s\otimes\rho_t\,dr\,ds\,dt = t_k,\end{aligned}\tag{8.7}$$

where $k\in C_{00}(G\times G)$ is given by $k(s,t)=\int\Delta(r)^2f(r)h(r^{-1}s,r^{-1}t)\,dr$. Thus $\delta_\rho\big(\rho(f)\big)\cdot t_h\in\rho(A)\otimes\rho(A)$. It follows by continuity that

$$\delta_\rho\big(\rho(a)\big)\cdot x\in\rho(A)\otimes\rho(A)$$

for all a in A and x in $\rho(A)\otimes\rho(A)$.

Since $\rho\otimes\rho$ is a faithful representation of $A\otimes A$, we may identify the idealiser of $\rho(A)\otimes\rho(A)$ in $\mathcal{L}(L^2(G\times G))$ with $M(A\otimes A)$ (by Proposition 2.3), and with this identification we have proved that $\delta_\rho(\rho(a)) \in M(A\otimes A)$ for all a in A. Now define $\delta: A \to M(A\otimes A)$ by $\delta(a) = \delta_\rho(\rho(a))$ $(a \in A)$.

Next, we show that $\delta: A \to M(A\otimes A)$ is nondegenerate. To see this, for each compact neighbourhood U of the identity in G, define $f_U = \chi_U / \int_G \chi_U$ (where χ_U is the characteristic function of U). With the compact neighbourhoods of e ordered by reverse inclusion, (f_U) is a directed net which is evidently in the unit ball of $L^1(G)$, and therefore in the unit ball of A. If $h \in C_{00}(G \times G)$ then by (8.7) $\delta(f_U) \cdot t_h = t_k$, where

$$k(s,t) = \int \Delta(r)^2 f_U(r) h(r^{-1}s, r^{-1}t)\, dr.$$

It follows from the uniform continuity of h that by taking U small enough we can make $\Delta(r)^2 h(r^{-1}s, r^{-1}t)$ close to $h(s,t)$ uniformly for r in U, and thereby ensure that k is close to h in the $L^1(G \times G)$ norm. Thus $\delta(f_U) \cdot t_h \xrightarrow{U} t_h$ in $A\otimes A$ (which we are identifying, as above, with $\rho(A)\otimes\rho(A)$. Since $\{t_h : h \in C_{00}(G \times G)\}$ is dense in $A\otimes A$ and $\{f_U\}$ is bounded, it follows that $\delta(f_U)c \xrightarrow{U} c$ for all c in $A\otimes A$. Therefore $\delta(A) \cdot A\otimes A$ is dense in $A\otimes A$. That is, δ is nondegenerate.

The uniqueness of the extension of δ to a mapping from $M(A)$ to $\mathcal{L}(L^2(G \times G))$ (Proposition 2.1) shows that this extension is given by the same formula

$$\delta(x) = w^*(1\otimes\rho(x))w \qquad (x \in M(A)) \tag{8.8}$$

(where now of course ρ refers to the extension of the original ρ to a representation of $M(A)$). In particular,

$$\delta(u_r) = u_r \otimes u_r \qquad (r \in G). \tag{8.9}$$

Since the unit ball of $D(G)$ is strictly dense in the unit ball of $M(A)$ (Proposition A1 in the Appendix), we can describe δ as the unique element of $\mathrm{Mor}(A, A\otimes A)$ satisfying (8.9).

We have yet to show that σ satisfies the conditions (i), (ii) in the definition of a comultiplication. For (i), notice that by (8.9)

$$(\iota\otimes\delta)\delta(u_r) = (\delta\otimes\iota)\delta(u_r) = u_r\otimes u_r\otimes u_r \qquad (r \in G).$$

Hence

$$(\iota\otimes\delta)\delta(x) = (\delta\otimes\iota)\delta(x) \qquad (8.10)$$

for all x in $D(G)$. it follows from Proposition A1 that (8.10) holds for all x in $M(A)$.

For (ii), let $f, g \in C_{00}(G)$. Then

$$\begin{aligned}\delta_\rho(\rho(f))\cdot 1\otimes\rho(g) &= \iint f(r)g(s)\rho_r\otimes\rho_{rs}\,dr\,ds \\ &= \iint \Delta(r)f(r)g(r^{-1}s)\rho_r\otimes\rho_s\,dr\,ds = t_k,\end{aligned}$$

where $k(r,s) = \Delta(r)f(r)g(r^{-1}s)$. Since $k \in C_{00}(G\times G)$, it follows that $\delta_\rho(\rho(f))\cdot 1\otimes\rho(g) \in \rho(A)\otimes\rho(A)$, so that $\delta(f)\cdot 1\otimes g \in A\otimes A$. By continuity, this holds for all f, g in A. The other condition in (ii) is proved similarly.

That concludes the construction of the C*-bialgebra $(C_r^*(G),\delta)$. The remainder of the chapter consists of an introduction to the notion of a unitary corepresentation of C*-bialgebra. To motivate this concept, we shall look at a particular example, namely the case of the function C*-algebra of a group. But first we need to make a general observation about the comultiplication on a C*-bialgebra.

Suppose that (A,δ) is a C*-bialgebra and that H is a Hilbert space. It follows from the results in Chapter 4 that there is a natural unitary equivalence between the Hilbert $(A\otimes A)$-modules $(H\otimes A)\otimes_\delta(A\otimes A)$ and $H\otimes A\otimes A$ given on simple tensors by

$$(\xi\otimes a)\dot\otimes(c\otimes d) \mapsto \xi\otimes(\delta(a)\cdot c\otimes d) \qquad (\xi\in H,\ a,c,d\in A).$$

Using this unitary equivalence, we can regard the induced morphism

$$\delta_*\colon \mathcal{L}(H\otimes A) \to \mathcal{L}((H\otimes A)\otimes_\delta(A\otimes A)) \qquad (8.11)$$

as a mapping from $\mathcal{L}(H\otimes A)$ to $\mathcal{L}(H\otimes A\otimes A)$. A routine verification shows that its action on simple tensors is given by

$$\delta_*(t\otimes a) = t\otimes\delta(a), \qquad (8.12)$$

where $t \in \mathcal{L}(H)$ and $a \in A$ (more accurately, a is the operation of left multiplication by a in $\mathcal{L}(A)$).

Now specialise to the case where $A = C_0(G)$ for some locally compact group G and δ is given by (8.5). Recall from Chapter 3 that there are canonical equivalences

$$H \otimes A \approx C_0(G, H), \quad \mathcal{L}(H \otimes A) \cong C_b^{\mathrm{str}}(G, \mathcal{L}(H)).$$

Similarly,

$$H \otimes A \otimes A \approx C_0(G \times G, H), \quad \mathcal{L}(H \otimes A \otimes A) \cong C_b^{\mathrm{str}}(G \times G, \mathcal{L}(H)).$$

Thus in this case we can regard δ_* as a mapping from $C_b^{\mathrm{str}}(G, \mathcal{L}(H))$ to $C_b^{\mathrm{str}}(G \times G, \mathcal{L}(H))$, and δ_* is given by the formula

$$(\delta_* x)(s,t) = x(st) \qquad (x \in C_b^{\mathrm{str}}(G, \mathcal{L}(H)),\ s,t \in G). \tag{8.13}$$

In fact, if x is a 'rank-one' function of the form $x: s \mapsto f(s)\xi \ \ (s \in G)$, for some $f \in C_0(G)$, $\xi \in H$, then (8.13) follows from (8.5) and (8.12), and the general case follows by linearity and strict continuity.

Suppose that $u: G \to \mathcal{L}(H)$ is a (strongly continuous) unitary representation of G on H. We want to describe u in terms of the group function algebra $A = C_0(G)$. Since the map $s \mapsto u_s$ is a strong*-continuous, unitary-valued function from G to $\mathcal{L}(H)$, it corresponds (under the unitary equivalence in the previous paragraph) to a unitary element, which we still call u, of $\mathcal{L}(H \otimes A)$. Since u is a representation of G, it has the multiplicative property $u_{st} = u_s u_t \ \ (s,t \in G)$, and we can describe this in the setting of the function algebra as follows. Under the identification $H \otimes A \otimes A \approx C_0(G \times G, H)$, and using the leg notation introduced earlier in this chapter, the operator $u_{12} \in \mathcal{L}(H \otimes A \otimes A)$ corresponds to the function $(s,t) \mapsto u_s$; and u_{13} corresponds to the function $(s,t) \mapsto u_t$. By (8.13) the operator $\delta_* u$ corresponds to the function $(s,t) \mapsto u_{st}$. Thus the multiplicative property of u can be expressed in the form $\delta_* u = u_{12} u_{13}$.

For a C*-bialgebra (A, δ), we define a *unitary corepresentation* of A on the Hilbert space H to be a unitary element u of $\mathcal{L}(H \otimes A)$ such that $\delta_* u = u_{12} u_{13}$. (In this equation, the right-hand side is a product of two elements of $\mathcal{L}(H \otimes A \otimes A)$, and the left-hand side is an operator in $\mathcal{L}(H \otimes A) \otimes_\delta (A \otimes A)$ which we identify with $\mathcal{L}(H \otimes A \otimes A)$ as above.)

Another approach to the definition of a unitary corepresentation, which is in some ways more useful, goes as follows. Form the tensor product of the

$*$-homomorphisms $\iota \in \mathrm{Mor}(\mathcal{K}(H), \mathcal{K}(H))$ and $\delta \in \mathrm{Mor}(A, A\otimes A)$ (where ι is the identity map). Identifying A with $\mathcal{K}(A)$ as usual, and using properties of the exterior tensor product from Chapter 4, we have

$$\begin{aligned}\iota\otimes\delta &\in \mathrm{Mor}\big(\mathcal{K}(H)\otimes\mathcal{K}(A), \mathcal{K}(H)\otimes\mathcal{K}(A)\otimes\mathcal{K}(A)\big)\\ &\cong \mathrm{Mor}\big(\mathcal{K}(H\otimes A), \mathcal{K}(H\otimes A\otimes A)\big).\end{aligned}$$

Since ι and δ are nondegenerate, so is $\iota\otimes\delta$, which therefore extends in the usual way to a $*$-homomorphism between the multiplier algebras:

$$\iota\otimes\delta\colon \mathcal{L}(H\otimes A) \longrightarrow \mathcal{L}(H\otimes A\otimes A).$$

But the equation (8.12) tells us that (modulo the natural equivalence $H\otimes A\otimes A \approx (H\otimes A)\otimes_\delta(A\otimes A)$) $\iota\otimes\delta$ agrees with δ_* on $\mathcal{K}(H\otimes A)$. Since both $\iota\otimes\delta$ and δ_* are strictly continuous on the unit ball, they therefore agree on $\mathcal{L}(H\otimes A)$; and we can rewrite the definition of a unitary corepresentation in the form

$$(\iota\otimes\delta)u = u_{12}u_{13}. \tag{8.14}$$

The formulation (8.14) has the advantage that both sides represent operators on $H\otimes A\otimes A$ (without having to make use of any canonical unitary equivalences).

The definition of a unitary corepresentation in the form (8.14) has another advantage: it extends without any effort to the case where the Hilbert space H is replaced by a Hilbert C*-module. In fact, let C be a C*-algebra and let E be a Hilbert C-module. A unitary element u of $\mathcal{L}_{C\otimes A}(E\otimes A)$ is said to be a *unitary corepresentation of* A *on* E if it satisfies (8.14), where now both sides of (8.14) represent elements of $\mathcal{L}_{C\otimes A\otimes A}(E\otimes A\otimes A)$.

If u is a unitary corepresentation of A on H, and $\phi \in A^*$, then we can form the right slice $(\iota\otimes\phi)u \in \mathcal{L}(H)$. For ϕ, ψ in A^* we have by (8.3), (8.4) and Proposition 8.5,

$$\begin{aligned}(\iota\otimes\phi)u\cdot(\iota\otimes\psi)u &= (\iota\otimes\phi\otimes\psi)u_{12}u_{13}\\ &= (\iota\otimes\phi\otimes\psi)\delta_* u\\ &= \big(\iota\otimes(\phi\otimes\psi)\delta\big)u\\ &= \big(\iota\otimes(\phi\times\psi)\big)u.\end{aligned}$$

Thus the corepresentation u defines a homomorphism $\phi \mapsto (\iota\otimes\phi)u$ from the algebra A^* to $\mathcal{L}(H)$. (The same argument will work for a corepresentation on a Hilbert C-module E if we replace δ_* by $\iota\otimes\delta$ in the second line of the above displayed equations.)

In the case where $A = C_0(G)$ and u is a unitary representation of G (which gives rise to a unitary corepresentation of A as described above), the dual space A^* is the space $M(G)$ of bounded complex Radon measures on G, the multiplication on A^* defined in Proposition 8.5 is the convolution product on $M(G)$ and the map $\phi \mapsto (\iota\otimes\phi)u$ is the representation of $M(G)$ associated with u ([Dix 2], Proposition 13.3.1).

If $u \in \mathcal{L}(H\otimes A)$ is a unitary corepresentation of A then we can also form the left slice $(\omega\otimes\iota)u \in M(A)$, where $\omega \in \mathcal{L}(H)_*$. We call such elements of $M(A)$ *coefficients* of u. In the case where $A = C_0(G)$ and ω is the vector functional $\omega_{\xi\eta}$, given by

$$\omega_{\xi\eta}(t) = \langle \xi, t\eta \rangle \qquad (t \in \mathcal{L}(H),\ \xi, \eta \in H),$$

it is easy to see that $(\omega_{\xi\eta}\otimes\iota)u$ corresponds to the function $s \mapsto \langle \xi, u_s\eta \rangle$ in $C_b(G) \cong M(A)$.

References for Chapter 8: Most of these ideas are implicit or explicit in the introductory sections of [BaaSka]. The material on group convolution C*-bialgebras draws very heavily on [Lan 2].

Appendix

The reduced C*-algebra of a locally compact group

We assume knowledge of the theory of locally compact groups up to the existence of the Haar integral and the modular function, as contained in [HewRos] for example. In line with our convention that inner products are linear in the second variable, it turns out to be more convenient to use the right Haar measure rather than the more usual left Haar measure.

Let G be a locally compact group with right Haar measure dt and modular function Δ. For f in $L^1(G)$ and s in G, we have

$$\int f(ts)\,dt = \int f(t)\,dt, \tag{A1}$$

$$\int f(st)\,dt = \Delta(s)\int f(t)\,dt, \tag{A2}$$

$$\int f(t^{-1})\,dt = \int \Delta(t)f(t)\,dt. \tag{A3}$$

(All integrals are over G unless otherwise specified.) The space $L^1(G)$ is a Banach algebra with convolution product (denoted by $\times$):

$$(f\times g)(s) = \int f(st^{-1})g(t)\,dt = \int \Delta(t)f(t)g(t^{-1}s)\,dt. \tag{A4}$$

It becomes a Banach $*$-algebra under the involution given by

$$f^*(s) = \Delta(s)\overline{f(s^{-1})}. \tag{A5}$$

For a function $f\colon G\to \mathbf{C}$ and $s\in G$ we write f_s, ${}_sf$ for the functions given by

$$f_s(t) = f(ts), \quad {}_sf(t) = f(st) \qquad (t\in G). \tag{A6}$$

If $f\in L^1(G)$ then $\|f_s\|_1 = \|f\|_1 = \Delta(s^{-1})\|{}_sf\|_1$, and if $x\in L^2(G)$ then

$$\|x_s\|_2 = \|x\|_2 = \Delta(s)^{-\frac{1}{2}}\|{}_sx\|_2 = \Delta(s)^{\frac{1}{2}}\|{}_{s^{-1}}x\|_2 \tag{A7}$$

by (A1) and (A2) (where the norms are the L^1 norm for f and the L^2 norm for x).

For f in $L^1(G)$ we write $\check{f}$ for the function given by $\check{f}(t) = f(t^{-1})$ $(t\in G)$. This need not be in $L^1(G)$, but the (pointwise) product $\Delta\check{f}$ is in $L^1(G)$ and in fact $\|\Delta\check{f}\|_1 = \|f\|_1$ (by (A3)).

The left and right regular representations of G are defined by

$$\lambda_s(x) = \Delta(s)^{\frac{1}{2}}\,{}_{s^{-1}}x, \quad \rho_s(x) = x_s \qquad (x \in L^2(G),\ s \in G). \tag{A8}$$

The maps $s \mapsto \lambda_s$, $s \mapsto \rho_s$ are unitary representations of G on $L^2(G)$. By the general theory of group representations ([Dix 2], Chapter 13), they have integrated forms which are norm-reducing $*$-representations of $L^1(G)$. What this means is that if $f \in L^1(G)$ and $x, y \in L^2(G)$ then

$$\langle x, \lambda(f)y\rangle = \int f(r)\langle x, \lambda_r y\rangle\, dr, \tag{A9}$$

$$\langle x, \rho(f)y\rangle = \int f(r)\langle x, \rho_r y\rangle\, dr. \tag{A10}$$

We write these relations in the form

$$\lambda(f) = \int f(r)\lambda_r\, dr, \quad \rho(f) = \int f(r)\rho_r\, dr, \tag{A11}$$

remembering that these vector integrals exist in the weak operator topology; that is to say, (A11) is just an abbreviation for (A9) and (A10). It is evident from (A9) and (A10) that $\|\lambda(f)\| \leqslant \|f\|_1$, $\|\rho(f)\| \leqslant \|f\|_1$. In terms of the convolution operation,

$$\lambda(f)x = (\Delta^{-\frac{1}{2}}f) \times x, \quad \rho(f)x = x \times (\Delta\check{f}) \qquad (f \in L^1(G),\ x \in L^2(G)). \tag{A12}$$

If we define $u\colon L^2(G) \to L^2(G)$ by

$$ux(t) = \Delta(t)^{\frac{1}{2}}x(t^{-1}) \qquad (x \in L^2(G),\ t \in G) \tag{A13}$$

then u is a selfadjoint unitary operator. By a straightforward verification, $u\rho(f)u = \lambda(f)$ for all $f \in L^1(G)$, so λ and ρ are unitarily equivalent. We define the *reduced C^*-norm* on $L^1(G)$ by

$$\|f\| = \|\lambda(f)\| = \|\rho(f)\|; \tag{A14}$$

and the completion of $L^1(G)$ with respect to this norm is called the *reduced C^*-algebra*, or the *convolution C^*-algebra*, of G, denoted by $C_r^*(G)$. Note that λ and ρ extend to faithful $*$-representations (which we still denote by λ and ρ) of $C_r^*(G)$ on $L^2(G)$.

Further straightforward verifications establish that if $f \in L^1(G)$ and $s \in G$ then

$$\rho_s \rho(f) = \Delta(s)\rho({}_{s^{-1}}f), \quad \rho(f)\rho_s = \rho(f_{s^{-1}}). \tag{A15}$$

Thus ρ_s is a multiplier of $\rho(C^*_r(G))$, and by Proposition 2.3 ρ_s is the image under ρ of a unique element u_s of $M(C^*_r(G))$. The closed linear span of $\{u_s : s \in G\}$ is a C*-subalgebra of $M(C^*_r(G))$ which we shall call $D(G)$. (We could also form a C*-subalgebra of $M(C^*_r(G))$ from the left multipliers λ_s, but this would give the same subalgebra $D(G)$ because of the unitary equivalence of λ and ρ. In fact, $\lambda_s = \lambda(u_s)$.) We shall use the convolution symbol $\times$ for multiplication in $M(C^*_r(G))$. Thus

$$u_s \times f = \Delta(s)_{s^{-1}}f, \quad f \times u_s = f_{s^{-1}} \qquad (f \in L^1(G),\ s \in G). \tag{A16}$$

PROPOSITION A1. *The C*-algebra $D(G)$ is strictly dense in $M(C^*_r(G))$. The unit ball of $D(G)$ is strictly dense in the unit ball of $M(C^*_r(G))$.*

Proof. We prove the first statement, and the second statement then follows by Proposition 1.4. Since the convolution algebra $C_{00}(G)$ of all continuous functions on G with compact support is norm-dense in $C^*_r(G)$, which in turn is strictly dense in its multiplier algebra, it suffices to prove that each element of $C_{00}(G)$ is in the strict closure of $D(G)$.

Let $f \in C_{00}(G)$, and denote by $\operatorname{supp}(f)$ the (compact) support of f. For each open neighbourhood U of the identity e in G, the translates $\{sU : s \in \operatorname{supp}(f)\}$ cover $\operatorname{supp}(f)$. Choose a finite subcover $\{s_1U, \ldots, s_nU\}$ and let

$$S_j = \Bigl(s_jU \setminus \bigcup_{i=1}^{j-1} s_iU\Bigr) \cap \operatorname{supp}(f) \qquad (1 \leqslant j \leqslant n).$$

Then $\{S_1, \ldots, S_n\}$ is a partition of $\operatorname{supp}(f)$ into Borel subsets. Let

$$h_U = \sum_{j=1}^{n} \int_{S_j} f(t)\,dt \cdot u_{s_j} \in D(G).$$

Since u_{s_j} is unitary it is clear that $\|h_U\| \leqslant \sum_j \int_{S_j} |f(t)|\,dt = \|f\|_1$. So as U runs through the open neighbourhoods of e, $\{h_U\}$ is bounded by the $L^1(G)$ norm of f. With the set of open neighbourhoods of e ordered by reverse inclusion, we shall show that $h_U \to f$ strictly, or in other words

$$(h_U - f) \times g \to 0, \quad (h_U - f)^* \times g \to 0$$

in the $C^*_r(G)$ norm, for every g in $C^*_r(G)$. Since $\{h_U\}$ is bounded, it will in fact suffice to prove this just when $g \in C_{00}(G)$. We shall show that, in this case, the above convergence occurs for the $L^1(G)$ norm (which implies convergence in the $C^*_r(G)$ norm).

For r in G,

$$\begin{aligned}((h_U - f) \times g)(r) &= \sum_j \Big(\int_{S_j} f(t)\,dt\,\Delta(s_j)g(s_j^{-1}r) - \int_G \Delta(t)f(t)g(t^{-1}r)\,dt\Big)\\ &= \sum_j \int_{S_j} f(t)\big[\Delta(s^j)g(s_j^{-1}r) - \Delta(t)g(t^{-1}r)\big]\,dt.\end{aligned}$$

(The first line of the above calculation uses (A16) and the second of the two formulas (A4), and the second line uses the fact that $\{S_j\}$ is a partition of $\mathrm{supp}(f)$.) Now observe that, since $S_j \subseteq s_jU$ and g is uniformly continuous, by choosing U small enough we can ensure that $g(t^{-1}r)$ is close to $g(s_j^{-1}r)$ uniformly for t in S_j and r in G. Also, we can ensure that $\Delta(t)$ is uniformly close to $\Delta(s_j)$ $(t \in S_j)$. Combining these observations and noting that Δ is bounded on $\mathrm{supp}(f)$, we see that by choosing U small enough we can make $\Delta(s_j)g(s_j^{-1}r) - \Delta(t)g(t^{-1}r)$ uniformly small for $(t, r) \in S_j \times G$. More precisely, let $\varepsilon > 0$ and let M be the Haar measure of the compact set $\mathrm{supp}(f) \cdot \mathrm{supp}(g)$. Then there is a neighbourhood U_0 of e such that, whenever $U \subseteq U_0$,

$$\big|\Delta(s_j)g(s_j^{-1}r) - \Delta(t)g(t^{-1}r)\big| < \frac{\varepsilon}{\|f\|_1 M}$$

for all t in S_j $(1 \leqslant j \leqslant n)$ and all r in G. It follows that, for $U \subseteq U_0$,

$$\begin{aligned}\|(h_U - f) \times g\|_1 &\leqslant \sum_j \int_{S_j}\int_G |f(t)|\big|\Delta(s_j)g(s_j^{-1}r) - \Delta(t)g(t^{-1}r)\big|\,dr\,dt\\ &\leqslant \sum_j \int_{S_j} \frac{|f(t)|\varepsilon}{\|f\|_1}\,dt = \varepsilon.\end{aligned}$$

So we have proved that $\|(h_U - f) \times g\|_1 \xrightarrow{U} 0$; a similar argument, starting with the calculation

$$((h_U - f)^* \times g)(r) = \sum_j \int_{S_j} \overline{f(t)}\big[\Delta(s_j^{-1})g(s_jr) - \Delta(t^{-1})g(tr)\big]\,dt,$$

shows that $\|(h_U - f)^* \times g\|_1 \xrightarrow{U} 0$.

Chapter 9

Unbounded operators

As we have seen, a unitary representation of a locally compact group G gives rise to a unitary corepresentation of the function C*-bialgebra $C_0(G)$. Many groups (for example, groups of matrices) are most conveniently specified in terms of some fundamental representation, and the corresponding C*-bialgebras of functions will then have a fundamental corepresentation. Unfortunately, if the group is not compact then the fundamental representation is not usually unitary or even bounded. For a concrete example, the (double covering of the) group of euclidean motions of the plane is given by

$$E(2) = \left\{ s = \begin{pmatrix} \alpha & \beta \\ 0 & \bar{\alpha} \end{pmatrix} \in M_2(\mathbf{C}) : |\alpha| = 1 \right\}.$$

The fundamental representation of $G = E(2)$ is that on $\mathbf{C}^2$ displayed by the above equation. The associated corepresentation of $C_0(G)$ should be the operator in $\mathcal{L}(\mathbf{C}^2 \otimes C_0(G))$ that corresponds to the function $v: s \mapsto \left(\begin{smallmatrix} \alpha & \beta \\ 0 & \bar{\alpha} \end{smallmatrix}\right)$ in $C_b^{\text{str}}(G, \mathcal{L}(\mathbf{C}^2))$ under the usual equivalence between these two spaces. But this does not make sense, because the function v is not bounded (β is an unbounded function of s). Thus v, considered as an operator on $\mathbf{C}^2 \otimes C_0(G)$, not only fails to be unitary, but is not even bounded or everywhere defined.

This motivates us to study unbounded, densely defined operators on Hilbert C*-modules. We shall attempt to develop a theory of such operators by analogy with the theory of unbounded operators on Hilbert spaces, and some acquaintance with that theory is a prerequisite for understanding what follows. In this theory (as presented for example in [KadRin] or [Rud]), it is a fundamental property that if t is a closed densely defined operator on

H with a densely defined adjoint t^* then $1+t^*t$ is the inverse of a bounded operator on H. We shall see later in the chapter that this need not be the case for operators on Hilbert C*-modules, and it is necessary to add an additional condition, called regularity, in order to achieve a viable theory.

Let A be a C*-algebra and let E, F be Hilbert A-modules. For convenience, we shall use the notation $t: E \to F$ to indicate that t is an A-linear operator whose domain $D(t)$ is a dense submodule of E (not necessarily the whole of E as the notation would appear to suggest) and whose range is in F. Given $s, t: E \to F$, we write $s \subseteq t$ if $D(s) \subseteq D(t)$ and $sx = tx$ for all x in $D(s)$; and $s = t$ means that $s \subseteq t$ and $t \subseteq s$. Sums and products of unbounded operators are defined just as for Hilbert space operators. That is to say,

$$D(s+t) = D(s) \cap D(t), \quad (s+t)x = sx + tx \qquad (x \in D(s+t));$$

and if $r: F \to G$ then

$$D(rt) = \{x \in D(t): tx \in D(r)\}, \quad (rt)x = r(tx) \qquad (x \in D(rt)).$$

As in the Hilbert space case, there is no guarantee that sums or products of densely defined unbounded operators will be densely defined.

Given $t: E \to F$, we define a submodule $D(t^*)$ of F by

$$D(t^*) = \{y \in F: \text{there exists } z \text{ in } E \text{ with } \langle tx, y\rangle = \langle x, z\rangle \text{ for all } x \in D(t)\}. \tag{9.1}$$

For y in $D(t^*)$ the element z in (9.1) is unique and is written $z = t^*y$. This defines an A-linear map $t^*: D(t^*) \to E$ satisfying

$$\langle x, t^*y\rangle = \langle tx, y\rangle \qquad (x \in D(t),\ y \in D(t^*)).$$

We define the graph of t to be the submodule $G(t)$ of $E \oplus F$ given by

$$G(t) = \{(x, tx): x \in D(t)\},$$

and we define a unitary element v of $\mathcal{L}(E \oplus F, F \oplus E)$ by $v(x, y) = (y, -x)$. As in the case of Hilbert space operators, it is quite easy to verify that $G(t^*) = vG(t)^\perp$. It follows that t^* is closed (which just means that its graph is a closed submodule of $F \oplus E$). If E, F were Hilbert spaces and t

were closed then we would have $G(t) \oplus vG(t^*) = E \oplus F$, and the standard theory of closed linear operators on Hilbert spaces is built on this fact. But, as we know, closed submodules of Hilbert C*-modules need not be complemented, and this is the cause of all the difficulties that follow. In the Hilbert space case, a densely defined closed operator has a densely defined adjoint. We would not expect that to hold for operators on Hilbert C*-modules, since even bounded operators need not be adjointable. So the property of having a densely defined adjoint has to be built in as part of the following fundamental definition: a *regular* operator from E to F is a densely defined closed A-linear map $t: D(t) \to F$ such that t^* is densely defined and $1 + t^*t$ has dense range.

LEMMA 9.1. *If t is regular then t^*t is densely defined.*

Proof. Let $D(t^*t)$ be the domain of t^*t, so by definition

$$D(t^*t) = \{x \in D(t): tx \in D(t^*)\}.$$

Let M be the closure of $D(t^*t)$ and let L be the range of $1+t^*t$. We wish to show that $M = E$. By hypothesis L is dense in E (and we shall eventually show that L must be the whole of E).

For x in $D(t^*t)$ we have

$$\langle x, (1 + t^*t)x \rangle = \langle x, x \rangle + \langle tx, tx \rangle \geqslant \langle x, x \rangle,$$

and so $\|(1+t^*t)x\| \geqslant \|x\|$. Thus $1+t^*t$ is injective, and its densely defined inverse is norm-reducing. So $(1 + t^*t)^{-1}: L \to E$ extends to a bounded operator on E which by Lemma 4.1 is a positive element of $\mathcal{L}(E)$. Denote by q its square root in $\mathcal{L}(E)$. Then $0 \leqslant q \leqslant 1$. Also, $\operatorname{ran}(q) \subseteq M$ by Proposition 3.7, since clearly $\operatorname{ran}(q^2) \subseteq M$.

For y in L, we have $q^2y \in D(t^*t)$, and so

$$\begin{aligned} \langle tq^2y, tq^2y \rangle &= \langle q^2y, t^*tq^2y \rangle \\ &\leqslant \langle q^2y, (1 + t^*t)q^2y \rangle = \langle q^2y, y \rangle \leqslant \langle y, y \rangle. \end{aligned} \tag{9.2}$$

Thus $\|tq^2y\| \leqslant \|y\|$ for all y in the dense submodule L, and so tq^2 has an extension to a bounded map r from E to F. In fact, no extension is necessary, because $\operatorname{ran}(q^2) \subseteq D(t)$. To see this, let $x \in E$. Since L is dense in E, we can choose a sequence (z_n) in $D(t^*t)$ with $(1+t^*t)z_n \to x$ as $n \to \infty$. Since q^2 is continuous, $z_n \to q^2x$; and since r is continuous, $tz_n \to rx$. Since t is closed, it follows that $q^2x \in D(t)$ and $tq^2x = rx$.

Let $x \in E$, $y \in D(t^*)$. Since $q^2x \in D(t)$, we have

$$\langle x, q^2t^*y\rangle = \langle q^2x, t^*y\rangle = \langle tq^2x, y\rangle = \langle rx, y\rangle$$

and so $\|q^2t^*y\| \leqslant \|y\|$. Thus $q^2t^*: D(t^*) \to E$ extends to a bounded map from F to E which is an adjoint for r. Thus $r \in \mathcal{L}(E,F)$, and $\mathrm{ran}(r^*) \subseteq M$.

For y in L and z in $D(t)$,

$$\begin{aligned}\langle y, (r^*t+q^2)z\rangle &= \langle tq^2y, tz\rangle + \langle q^2y, z\rangle \\ &= \langle (t^*t+1)q^2y, z\rangle = \langle y, z\rangle\end{aligned}$$

(recall that $q^2y \in D(t^*t)$, so that $tq^2y \in D(t^*)$). Since L is dense in E, this shows that $z = (r^*t+q^2)z \in M$. Hence $D(t) \subseteq M$ and so $M = E$.

Notice that $\mathrm{ran}(q^2) \supseteq D(1+t^*t) = D(t^*t)$, so that q^2 (and therefore also q) has dense range.

As in the case of Hilbert space operators, we say that a submodule K of the domain of a closed operator t is a *core* for t if t is the closure of its restriction to K.

LEMMA 9.2. *If t is regular then $D(t^*t)$ is a core for t.*

Proof. We shall continue to use the notation introduced in the proof of Lemma 9.1. For y in L, we have by (9.2)

$$\langle tq^2y, tq^2y\rangle \leqslant \langle q^2y, y\rangle = \langle qy, qy\rangle.$$

Therefore the map $f: qy \mapsto tq^2y$ extends from the dense submodule $q(L)$ to a norm-reducing map from E to F. Just as in the case of r, in the proof of Lemma 9.1, no extension is actually needed: the fact that t is closed implies that $\mathrm{ran}(q) \subseteq D(t)$, so that $f = tq$. Also, the map $qt^*: D(t^*) \to E$ extends to a bounded map which is an adjoint for f, so that $f \in \mathcal{L}(E,F)$. It is clear that $r = fq$. So $r^* = qf^*$.

By the last paragraph of the proof of Lemma 9.1, if $y \in D(t)$ then

$$y = (r^*t+q^2)y = q(f^*t+q)y \in \mathrm{ran}(q).$$

Thus we have shown that $\operatorname{ran}(q) = D(t)$, and therefore $t = fq^{-1}$. So given y in $D(t)$ we can write $y = qx$ (for some x in E). Choose a sequence (z_n) in L with $qz_n \to x$ as $n \to \infty$ (this is possible since $q(L)$ is dense in E). We then have

$$q^2 z_n \to y,$$
$$tq^2 z_n = fqz_n \to fx = tqx = ty.$$

Hence $(q^2 z_n, tq^2 z_n) \to (y, ty)$ in $G(t)$. But $q^2 z_n \in D(t^*t)$ since $z_n \in L$, and so we have shown that $D(t^*t)$ is a core for t.

With these preparations, we are ready for the main theorem of this chapter, which states that if $t\colon E \to F$ is a regular operator then $G(t)$ is a complemented submodule of $E \oplus F$.

THEOREM 9.3. *Let E, F be Hilbert A-modules, let t be a regular operator from $D(t) \subseteq E$ to F, and let $v \in \mathcal{L}(E \oplus F, F \oplus E)$ be the unitary operator given by $v(x,y) = (y, -x)$ $(x \in E,\ y \in F)$. Then $G(t) \oplus vG(t^*) = E \oplus F$.*

Proof. We have already observed that the closed submodules $G(t)$ and $vG(t^*)$ of $E \oplus F$ are orthogonal. We shall show that the sum of these submodules is the whole of $E \oplus F$ by constructing a projection p in $\mathcal{L}(E \oplus F)$ such that $\operatorname{ran}(p) \subseteq G(t)$ and $\ker(p) \subseteq vG(t^*)$. We continue to use the notation developed in the proofs of Lemmas 9.1 and 9.2.

Let $x \in D(t^*t)$ and $y \in E$. Then $q^2 y \in D(t)$, since we know that $\operatorname{ran}(q) = D(t)$. So

$$\langle x, y\rangle = \langle q^2(1 + t^*t)x, y\rangle = \langle x, q^2 y\rangle + \langle tx, tq^2 y\rangle,$$

and therefore

$$\langle tx, tq^2 y\rangle = \langle x, (1 - q^2)y\rangle. \tag{9.3}$$

Since $D(t^*t)$ is a core for t, (9.3) holds for all x in $D(t)$ (and all y in E). Therefore $tq^2 y \in D(t^*)$ and $t^*tq^2 y = (1 - q^2)y$. Hence $(1 + t^*t)q^2 = 1$. This shows (as we promised in the proof of Lemma 9.1) that L is the whole of E, and also that $\operatorname{ran}(q^2) = D(t^*t)$. It also follows that q^2 is equal to $(1 + t^*t)^{-1}$ (rather than an extension of it, which was how we originally constructed q^2); and we can write $q = (1 + t^*t)^{-\frac{1}{2}}$.

At this stage, it may help if we summarise the properties of the operators $q \in \mathcal{L}(E)$, $f \in \mathcal{L}(E, F)$ that we have constructed from t. These are:

$$(1 + t^*t)q^2 = 1, \qquad q^2(1 + t^*t) \subseteq 1, \tag{9.4}$$

$$tq = f, \qquad qt^* \subseteq f^*. \tag{9.5}$$

There is a useful relation between q and f, which we obtain as follows. Let $x \in E$. Then $fqx = tq^2x \in D(t^*)$, and so

$$f^*fqx = qt^*tq^2x = q(1 - q^2)x = (1 - q^2)qx.$$

Since ran(q) is dense in E, it follows that $f^*f = 1 - q^2$, and therefore that

$$q = (1 - f^*f)^{\frac{1}{2}}. \tag{9.6}$$

Now define $b \in \mathcal{L}(E, E \oplus F)$ by $bx = (qx, fx) \quad (x \in E)$, or in matrix terms $b = \binom{q}{f}$. Then

$$b^*b = (q \quad f^*)\begin{pmatrix} q \\ f \end{pmatrix} = q^2 + f^*f = 1.$$

Thus b is an isometry, and it follows that $p = bb^*$ is a projection in $\mathcal{L}(E \oplus F)$, with matrix given by

$$p = \begin{pmatrix} q^2 & qf^* \\ fq & ff^* \end{pmatrix}, \quad 1 - p = \begin{pmatrix} f^*f & -qf^* \\ -fq & 1 - ff^* \end{pmatrix}. \tag{9.7}$$

Since ran$(b) \subseteq G(t)$ by (9.5), we must have ran$(p) \subseteq G(t)$. We shall complete the proof of the theorem by showing that ran$(1 - p) \subseteq vG(t^*)$. To see this, note that

$$(1 - p)(x, y) = (f^*fx, -fqx) + (-qf^*y, (1 - ff^*)y) \qquad (x \in E,\ y \in F),$$

so it will be sufficient to prove that

$$(fqx, f^*fx) \in G(t^*) \qquad (x \in E), \tag{9.8}$$

$$((1 - ff^*)y, qf^*y) \in G(t^*) \qquad (y \in F). \tag{9.9}$$

For (9.8), we use (9.4) and (9.5) to get

$$f^*fx = (1 - q^2)x = t^*tq^2x = t^*(fqx).$$

For (9.9), suppose that $y \in D(t^*)$. Then

$$\begin{aligned} qf^*y = q^2t^*y &= (1 - t^*tq^2)t^*y \\ &= t^*(1 - tq^2t^*)y = t^*(1 - ff^*)y. \end{aligned}$$

Since $D(t^*)$ is dense in F and t^* is closed, it follows that

$$qf^*y = t^*(1 - ff^*)y$$

for all y in F, as required.

COROLLARY 9.4. *If t is regular then $t^{**} = t$.*

Proof. Since $G(t)$ is complemented, $G(t^{**}) = G(t)^{\perp\perp} = G(t)$.

PROPOSITION 9.5. *Suppose that $t\colon D(t) \to F$ is closed, where $D(t)$ is a dense submodule of E. Suppose also that t^* is densely defined, and that $G(t) \oplus vG(t^*) = E \oplus F$. Then t is regular.*

Proof. The projection p from $E \oplus F$ onto $G(t)$ (with kernel $vG(t^*)$) is a positive element of $\mathcal{L}(E \oplus F)$ and hence has a matrix representation of the form

$$p = \begin{pmatrix} a & b^* \\ b & d \end{pmatrix}, \tag{9.10}$$

where $0 \leqslant a \leqslant 1$ in $\mathcal{L}(E)$, $b \in \mathcal{L}(E, F)$ and $0 \leqslant d \leqslant 1$ in $\mathcal{L}(F)$. Since $\operatorname{ran}(p) = G(t)$ and $\operatorname{ran}(1 - p) = vG(t^*)$, it follows that

$$\begin{aligned} &\operatorname{ran}(a) \subseteq D(t), \quad \text{with } b = ta, \\ &\operatorname{ran}(b) \subseteq D(t^*), \quad \text{with } 1 - a = t^*b. \end{aligned}$$

Hence $\operatorname{ran}(a) \subseteq D(t^*t)$, with $1 - a = t^*ta$. Thus $(1 + t^*t)a = 1$, so that $1 + t^*t$ is surjective. Therefore t is regular.

COROLLARY 9.6. *Let $t\colon E \to F$ be a closed operator, and suppose that both t and t^* are densely defined. Then t is regular if and only if t^* is regular.*

Proof. Since $v^2 = -1$ and $G(t)$ is invariant under multiplication by the scalar -1, the condition $G(t) \oplus vG(t^*) = E \oplus F$ is symmetric in t and t^*.

But by Theorem 9.3 and Proposition 9.5, this condition is equivalent to the regularity of t.

As in the case of Hilbert space operators, we say that a densely defined A-linear mapping t between Hilbert A-modules is *symmetric* if $t \subseteq t^*$, and *selfadjoint* if $t = t^*$.

LEMMA 9.7. *If $t: E \to F$ is closed, densely defined and symmetric then the operators $t \pm i$ are injective and have closed range.*

Proof. For x in $D(t)$,

$$\begin{aligned} |(t+i)x|^2 &= \langle tx + ix, tx + ix\rangle \\ &= \langle tx, tx\rangle - i\langle x, tx\rangle + i\langle tx, x\rangle + \langle x, x\rangle \\ &= |tx|^2 + |x|^2 \quad \text{(since } t \text{ is symmetric)} \\ &= |(x, tx)|^2. \end{aligned}$$

So the map $(x, tx) \mapsto (t+i)x$ is isometric from $G(t)$ onto $\operatorname{ran}(t+i)$. Since t is closed, so is $\operatorname{ran}(t+i)$. The above calculation also shows that $t+i$ is injective. The results about $t-i$ are proved in the same way.

LEMMA 9.8. *Suppose that $t: E \to E$ is densely defined and selfadjoint. Then t is regular if and only if the operators $t \pm i$ are surjective.*

Proof. If t is regular then $1 + t^2$ has dense range, and in fact is surjective as we saw in the proof of Theorem 9.3. Thus $(t+i)(t-i)$ is surjective and hence so is $t+i$. Also, $(t-i)(t+i)$ is surjective and hence so is $t-i$.

Conversely, if $t+i$ and $t-i$ are surjective then so is $1+t^2 = (t+i)(t-i)$ and so t is regular.

PROPOSITION 9.9. *If $t: E \to F$ is regular then t^*t is selfadjoint and regular.*

Proof. By Lemma 9.1, $D(t^*t)$ is dense. It is evident that t^*t is symmetric, so that $(t^*t)^*$ is also densely defined.

Next, we show that t^*t is selfadjoint. It is straightforwardly verified that $(1 + t^*t)^* = 1 + (t^*t)^*$, so it will be enough to show that $1 + t^*t$ is selfadjoint. With x in E, y in $D\big((t^*t)^*\big) = D\big(1+(t^*t)^*\big)$ and q^2 as in (9.4), we have

$$\langle x, y\rangle = \langle (1+t^*t)q^2x, y\rangle = \langle x, q^2(1+t^*t)^*y\rangle.$$

Since this holds for all x in E, it follows that

$$y = q^2(1+t^*t)^*y \in \operatorname{ran}(q^2) = D(1+t^*t).$$

Thus $D\big((1+t^*t)^*\big) = D(1+t^*t)$, and since $1+t^*t$ is symmetric it follows that it must be selfadjoint.

Finally, we use Lemma 9.8 to show that t^*t is regular. Given x in E, let $y = q^2\big(1-(1-i)q^2\big)^{-1}x$. (Since $q^2 \geqslant 0$ in $\mathcal{L}(E)$, $1-(1-i)q^2$ is certainly invertible.) Then by (9.4)

$$\begin{aligned}(t^*t+i)y &= (t^*t+i)q^2\big(1-(1-i)q^2\big)^{-1}x \\ &= \big(1-(1-i)q^2\big)\big(1-(1-i)q^2\big)^{-1}x = x.\end{aligned}$$

Hence t^*t+i (and similarly t^*t-i) is surjective. This completes the proof.

We shall next construct a class of examples which will enable us to show that, contrary to what one might expect from the case of Hilbert space operators, an unbounded closed operator t on a Hilbert C*-module can fail to satisfy the regularity condition, even though t and t^* are both densely defined.

Let A be the commutative unital C*-algebra $C(X)$, where X is a compact Hausdorff space, and let E be the Hilbert A-module $H\otimes A$, where H is a Hilbert space. As we know, E can be identified with $C(X,H)$ and $\mathcal{L}(E)$ with $C_b^{str}(X,\mathcal{L}(H))$. We shall consider (unbounded) operators on E given by mappings of the form $\lambda \mapsto t_\lambda$, where t_λ is a selfadjoint (unbounded) operator on H, for all λ in X.

Suppose that, for each λ in X, t_λ is a selfadjoint operator on H with domain D_λ. Suppose also that there is a dense subspace D of H such that $D \subseteq D_\lambda$ for all λ, and that for ξ in D the map $\lambda \mapsto t_\lambda\xi$ is continuous from X to H. Define $D(t) \subseteq E$ by

$$D(t) = \big\{x \in C(X,H) : x(\lambda) \in D_\lambda \text{ for all } \lambda, \\ \text{and the map } \lambda \mapsto t_\lambda x(\lambda) \text{ is continuous}\big\}.$$

Notice that $D(\lambda)$ is a submodule of E; and that it is dense in E because it contains the dense submodule $D \otimes_{\text{alg}} A$ of $H \otimes A$. For x in $D(t)$ define tx in E by $(tx)(\lambda) = t_\lambda x(\lambda)$ $(\lambda \in X)$. Then t is evidently a symmetric densely defined operator on E. If $y \in D(t^*)$ then, for all x in $D(t)$ and for all λ in X,

$$\langle x(\lambda), (t^* y)(\lambda) \rangle = \langle t_\lambda x(\lambda), y(\lambda) \rangle.$$

Therefore $y(\lambda) \in D(t_\lambda^*)$ and $t_\lambda^*\big(y(\lambda)\big) = (t^* y)(\lambda)$. But t_λ is selfadjoint, so this shows that $y \in D(t)$ and $t^* y = ty$. Thus we have proved that t is selfadjoint.

With $t: E \to E$ constructed in this way from the field $\lambda \mapsto t_\lambda$ of selfadjoint operators on H, it is clear that, for each μ in X, the μth coordinate of $D(t)$ is a subspace of D_μ:

$$\big(D(t)\big)_\mu \subseteq D(t_\mu).$$

We are interested in knowing when this inclusion becomes an equality. In other words, when does $D(t)$ "fill out its coordinates"?

Suppose that t is regular. Then by Lemma 9.8 $(t+i)D(t) = E$. Hence, for each μ in X,

$$(t_\mu + i)(D(t))_\mu = H.$$

But t_μ is a selfadjoint operator on H, so by Lemma 9.7 $t_\mu + i$ is an injective mapping from D_μ onto H. If its restriction to $\big(D(t)\big)_\mu$ is surjective then this must mean that $\big(D(t)\big)_\mu$ is the whole of D_μ. To summarise, if t is regular then $\big(D(t)\big)_\mu = D_\mu$ for all μ in X.

We are now ready to construct our example of a non-regular operator. We take $X = [0,1]$, $H = L^2\big([0,1]\big)$. If $x \in E = H \otimes C(X)$ then x is given by a continuous function $\lambda \mapsto x_\lambda$ from $[0,1]$ to $L^2\big([0,1]\big)$, and we write the L^2-function x_λ as $\mu \mapsto x_\lambda(\mu)$ $(\mu \in [0,1])$. Let

$$\alpha_\lambda = \begin{cases} 1 & (\lambda = 0), \\ e^{i/\lambda} & (\lambda > 0). \end{cases}$$

For $0 \leqslant \lambda \leqslant 1$, define

$$D_\lambda = \big\{\xi \in L^2\big([0,1]\big) : \xi \text{ is absolutely continuous,} \\ \xi' \in L^2\big([0,1]\big) \text{ and } \xi(1) = \alpha_\lambda \xi(0)\big\},$$

and for ξ in D_λ define $t_\lambda\xi = i\xi'$. It is well known that $t_\lambda = t_\lambda^*$ ([Rud], pp. 331–332; [KadRin], Exercise 2.8.50). If

$$D = \{\xi \in L^2([0,1]) : \xi \text{ is absolutely continuous, } \xi' \in L^2([0,1]) \text{ and } \xi(1) = \xi(0) = 0\}$$

then D is dense in $L^2([0,1])$ and $D \subseteq D_\lambda$ for all λ in $[0,1]$. Thus all the conditions are satisfied for us to construct a densely defined selfadjoint operator t from the field of operators t_λ as in the preceding discussion. If this t were regular then the argument in the previous paragraph would show that $(D(t))_\lambda = D_\lambda$ for all λ in $[0,1]$. We shall prove that $(D(t))_0 \neq D_0$, so that t is not regular (that is, $1 + t^*t$ cannot have dense range).

Define 1 in E by $1 : \lambda \mapsto 1$, where the final 1 is the constant function 1 in $L^2([0,1])$. If $x : \lambda \mapsto x_\lambda$ is any element of $D(t)$, and $0 \leqslant \lambda \leqslant 1$, then

$$\begin{aligned} \langle 1, tx\rangle(\lambda) &= i\int_0^1 x'_\lambda(\mu)\, d\mu \\ &= i(x_\lambda(1) - x_\lambda(0)) \\ &= i(\alpha_\lambda - 1)x_\lambda(0). \end{aligned}$$

Since $\langle 1, tx\rangle$ is a continuous function of λ and $\alpha_\lambda - 1$ is discontinuous at $\lambda = 0$, we must have $x_0(0) = 0$. Hence $(D(t))_0 = D$, which is a proper subspace of D_0. This completes the proof that t is not regular.

In the next chapter, we shall introduce a powerful technique due to Woronowicz which is a key to obtaining many of the deeper properties of regular operators. To conclude this chapter, however, we return to the problem raised in the opening paragraph, where it was suggested that there is a need to extend the concept of a unitary corepresentation to that of an "unbounded corepresentation". Naively, one might hope that an unbounded corepresentation of a C*-bialgebra (A, δ) on a Hilbert space H should be an invertible regular operator v on $H \otimes A$ such that $\delta_* v = v_{12}v_{13}$. But since we are dealing with unbounded operators, we need to scrutinise every aspect of this proposed definition to see whether it makes sense.

First, the proposed definition requires v to be invertible. We say that a regular operator $t : E \to F$ is *invertible* if t and t^* both have dense range. It is then a routine exercise to check that t and t^* are injective (so that t^{-1},

$(t^*)^{-1}$ are well-defined, with domains $\operatorname{ran}(t)$, $\operatorname{ran}(t^*)$ respectively), that t^{-1}, $(t^*)^{-1}$ are regular (use Proposition 9.5 for this) and that $(t^{-1})^* = (t^*)^{-1}$.

Next, we want to define v_{12} to be the operator $v \otimes 1$ on $(H \otimes A) \otimes A$. It is not too hard to see how to define $v \otimes 1$ as a regular operator. We shall not discuss how to do this, since in the next chapter we shall address the much more difficult problem of forming the exterior tensor product of any two regular operators. (The operator $v \otimes 1$ can also be regarded as a special case of the construction in the following proposition.) Thus v_{12} and similarly v_{13} are defined as regular operators on $H \otimes A \otimes A$.

To form $\delta_* v$, we need to consider how regular unbounded operators relate to the interior tensor product construction. Let E be a Hilbert A-module, let F be a Hilbert B-module and suppose that $\alpha: A \to \mathcal{L}(F)$ is a $*$-homomorphism and that $t: D(t) \to E$ is a regular operator on E. Let D_0 be the linear span of the set $\{x \dot{\otimes} y : x \in D(t),\ y \in F\}$, so that D_0 is a dense submodule of $E \otimes_\alpha F$. Let t_1 be the map from D_0 to $E \otimes_\alpha F$ defined on simple tensors by $t_1 \cdot x \dot{\otimes} y = t x \dot{\otimes} y$. In the same way, define D_0^* and $t_2: D_0^* \to E \otimes_\alpha F$ starting with the operator t^*. For x in $D(t)$, z in $D(t^*)$ and y, w in F, we have

$$\begin{aligned} \langle t_1 \cdot x \dot{\otimes} y, z \dot{\otimes} w \rangle &= \langle t x \dot{\otimes} y, z \dot{\otimes} w \rangle \\ &= \big\langle y, \alpha\big(\langle t x, z \rangle\big) w \big\rangle \\ &= \big\langle y, \alpha\big(\langle x, t^* z \rangle\big) w \big\rangle = \langle x \dot{\otimes} y,\, t_2 \cdot z \dot{\otimes} w \rangle. \end{aligned}$$

By linearity, $\langle t_1 \xi, \eta \rangle = \langle \xi, t_2 \eta \rangle$ whenever $\xi \in D_0$, $\eta \in D_0^*$. Therefore $D_0^* \subseteq D(t_1^*)$ and $t_2 \subseteq t_1^*$. In particular, t_1^* is densely defined and so t_1 is closable. We write $\alpha_*(t)$ for the closure of t_1. It is clear that if $t \in \mathcal{L}(E)$ then this coincides with the previous definition of $\alpha_*(t)$, in Chapter 4.

PROPOSITION 9.10. *If t is a regular operator on E then the operator $\alpha_*(t)$ defined above is regular on $E \otimes_\alpha F$, and $\big(\alpha_*(t)\big)^* = \alpha_*(t^*)$.*

Proof. We have already shown that $\alpha_*(t^*) \subseteq \big(\alpha_*(t)\big)^*$. Write v_1 for the unitary operator $(x, y) \mapsto (y, -x)$ on $E \oplus E$, and v_2 for the unitary operator on $(E \otimes_\alpha F) \oplus (E \otimes_\alpha F)$ given by the same formula. If $x \in E$ then, since $G(t) \oplus v_1 G(t^*) = E \oplus E$, it follows that

$$(x, 0) = (y + t^* z, t y - z)$$

for some y in $D(t)$ and z in $D(t^*)$. For w in F, we have $y\dot{\otimes}w \in D(\alpha_*(t))$, $z\dot{\otimes}w \in D(\alpha_*(t^*))$, and

$$\begin{aligned}(x\dot{\otimes}w, 0) &= (y\dot{\otimes}w + \alpha_*(t^*)\cdot z\dot{\otimes}w,\ \alpha_*(t)\cdot y\dot{\otimes}w - z\dot{\otimes}w) \\ &\in G(\alpha_*(t)) \oplus v_2 G(\alpha_*(t^*)).\end{aligned}$$

Thus $(E\otimes_\alpha F) \oplus \{0\} \subseteq G(\alpha_*(t)) \oplus v_2G(\alpha_*(t^*))$. A similar argument shows that $\{0\} \oplus (E\otimes_\alpha F) \subseteq G(\alpha_*(t)) \oplus v_2G(\alpha_*(t^*))$. We conclude that $G(\alpha_*(t)) \oplus v_2G(\alpha_*(t^*))$ is the whole of $(E\otimes_\alpha F) \oplus (E\otimes_\alpha F)$. This shows that $\alpha_*(t^*) = (\alpha_*(t))^*$ and also (by Proposition 9.5) that $\alpha_*(t)$ is regular.

Returning to our proposed definition of an unbounded corepresentation as an invertible regular operator v on $H\otimes A$ such that

$$\delta_* v = v_{12}v_{13}, \tag{9.11}$$

we see that both sides of (9.11) have now been defined and that the operator on the left-hand side is necessarily regular. It has to be admitted that unbounded corepresentations have not yet appeared in the literature, and it remains to be seen whether the concept will prove useful.

References for Chapter 9: [BaaJul], [Hil], [Wor 5].

Chapter 10

The bounded transform, unbounded multipliers

In the previous chapter, we saw that to each regular operator $t: E \to F$ between Hilbert A-modules E and F there correspond bounded adjointable operators q, f satisfying (9.4) and (9.5). By the proof of Lemma 9.2 we can reconstruct t from q and f by the formula $t = fq^{-1}$. In fact, by (9.6) we can express q in terms of f and thereby reconstruct t from f alone (see (10.2) below). In this chapter we shall use the notation q_t and f_t in place of q and f, and we shall call f_t the *bounded transform* of t.

In [Wor 5] and [WorNap], Woronowicz (with Napiórkowski) uses the notation z_t for f_t and calls it the z-transform of t. He shows how the bounded transform can be used as the fundamental tool for developing the whole theory of regular operators. This chapter will be devoted to an exposition of some of the major results in the above two papers.

We can summarise the facts about the bounded transform that were established in Chapter 9 as follows. Given a regular operator $t: E \to F$, we have $f_t \in \mathcal{L}(E, F)$, $q_t \in \mathcal{L}(E)$ and

$$f_t = t(1 + t^*t)^{-\frac{1}{2}}, \tag{10.1}$$

$$t = f_t(1 - f_t^* f_t)^{-\frac{1}{2}}, \tag{10.2}$$

$$q_t = (1 - f_t^* f_t)^{\frac{1}{2}} = (1 + t^*t)^{-\frac{1}{2}}. \tag{10.3}$$

Furthermore,

$$0 \leqslant q_t \leqslant 1 \text{ in } \mathcal{L}(E), \quad \operatorname{ran}(q_t) = D(t), \tag{10.4}$$

$$\|f_t\| \leqslant 1 \text{ in } \mathcal{L}(E,F), \quad \overline{\operatorname{ran}(1-f_t^*f_t)} = E. \tag{10.5}$$

Our first main result is that (10.5) characterises those elements of $\mathcal{L}(E,F)$ that can arise as bounded transforms of regular operators. We need some preparatory results first.

Recall from Chapter 8 that if A is a C*-algebra then $M(A)_*$ denotes the bounded linear functionals on A that are strictly continuous on the unit ball. If $h \geqslant 0$ in $M(A)$ then we shall say that h is *strictly positive* mod A if $\rho(h) > 0$ for every state ρ in $M(A)_*$. Note that if A is nonunital then this is not the same as saying that h is strictly positive. Indeed, if h is a strictly positive element of A then h (considered as an element of $M(A)$) is strictly positive mod A; but h is not strictly positive as an element of $M(A)$ because there are states of $M(A)$ that annihilate A.

The proof of Lemma 6.1 applies virtually unchanged to show that a positive element h of $M(A)$ is strictly positive mod A if and only if hA is dense in A. The following lemma is a slight generalisation of Lemma 1.1.22 in [JenTho].

LEMMA 10.1. *If E is a Hilbert A-module and $h \geqslant 0$ in $\mathcal{L}(E)$ then the following conditions are equivalent:*

(i) *h is strictly positive* mod $\mathcal{K}(E)$,
(ii) $\operatorname{ran}(h)$ *is dense in E,*
(iii) *$h^{\frac{1}{n}} \to 1$ strictly as $n \to \infty$.*

Proof. (i) $\Rightarrow$ (ii): By the generalisation of Lemma 6.1 quoted above, if h is strictly positive mod $\mathcal{K}(E)$ then $h\mathcal{K}(E)$ is dense in $\mathcal{K}(E)$. Since $\mathcal{K}(E)$ acts nondegenerately on E, this implies that h has dense range.

(ii) $\Rightarrow$ (iii): If $x = hy \in \operatorname{ran}(h)$ then $h^{\frac{1}{n}}x = h^{1+\frac{1}{n}}y$. Since $\lambda^{1+\frac{1}{n}} \to \lambda$ uniformly on $\operatorname{sp}(h)$, it follows by functional calculus that $h^{\frac{1}{n}}x \to x$ as $n \to \infty$. If $\operatorname{ran}(h)$ is dense then by continuity (and boundedness of $(h^{\frac{1}{n}})$) we have $h^{\frac{1}{n}}x \to x$ for all x in E.

(iii) $\Rightarrow$ (ii): Given $n \geqslant 1$ and $\varepsilon > 0$, choose f in $C([0,1])$ with

$$\left|\lambda^{\frac{1}{n}} - \lambda f(\lambda)\right| \leqslant \varepsilon \qquad (0 \leqslant \lambda \leqslant 1)$$

(for example, $f(\lambda) = \varepsilon^{1-n}$ when $0 \leqslant \lambda \leqslant \varepsilon^n$, $f(\lambda) = \lambda^{\frac{1}{n}-1}$ when $\varepsilon^n \leqslant \lambda \leqslant 1$). For x in E, $\|h^{\frac{1}{n}}x - hf(h)x\| \leqslant \varepsilon\|x\|$. Since this can be done for any $\varepsilon > 0$, $h^{\frac{1}{n}}x \in \overline{\operatorname{ran}(h)}$. If $h^{\frac{1}{n}} \to 1$ strictly then $h^{\frac{1}{n}}x \to x$, so $x \in \overline{\operatorname{ran}(h)}$. Thus $\overline{\operatorname{ran}(h)} = E$.

(ii) $\Rightarrow$ (i): Suppose that h has dense range. Given x, y in E, choose a sequence (x_n) in E with $hx_n \to x$. Then by (1.6)

$$\theta_{x,y} = \lim_{n\to\infty} \theta_{hx_n,y} = \lim_{n\to\infty} h\theta_{x_n,y} \in \overline{h\mathcal{K}(E)}.$$

Therefore $\mathcal{K}(E) = \overline{h\mathcal{K}(E)}$. Using the generalised Lemma 6.1 again, we conclude that h is strictly positive mod $\mathcal{K}(E)$.

COROLLARY 10.2. *If $0 \leqslant h \leqslant k$ in $\mathcal{L}(E)$ and h has dense range then so does k.*

Proof. It is evident that if h is strictly positive mod $\mathcal{K}(E)$ then so is k.

LEMMA 10.3. *If E, F are Hilbert A-modules, $f \in \mathcal{L}(E,F)$ and $\|f\| \leqslant 1$ then $1 - f^*f$ has dense range if and only if $1 - ff^*$ has dense range.*

Proof. We need only prove the implication in one direction since the reverse implication will then follow on exchanging f and f^*.

Suppose that $\operatorname{ran}(1 - ff^*)$ is not dense in F. By Lemma 10.1 there is a state ρ in $\mathcal{L}(F)_*$ with $\rho(1 - ff^*) = 0$, so $\rho(ff^*) = 1$. By functional calculus on the commutative C*-algebra generated by ff^*, we must have $\rho((ff^*)^2) = 1$ too.

Define $\sigma \in \mathcal{L}(E)_*$ by $\sigma(t) = \rho(ftf^*)$ for $t \in \mathcal{L}(E)$. Then σ is a state, and $\sigma(f^*f) = 1$. So $\sigma(1 - f^*f) = 0$ and therefore $1 - f^*f$ is not strictly positive mod $\mathcal{K}(E)$. By Lemma 10.1, $\operatorname{ran}(1 - f^*f)$ is not dense in E.

Observe that if the operators $1 - f^*f$, $1 - ff^*$ have dense range then they are injective (since the kernel of a selfadjoint operator is contained in the orthogonal complement of its range), and therefore so are their square roots.

THEOREM 10.4. *Let E, F be Hilbert A-modules, let $\mathcal{T}$ denote the set of regular operators from E to F and let*

$$\mathcal{F} = \{f \in \mathcal{L}(E,F) : \|f\| \leqslant 1 \text{ and } \operatorname{ran}(1 - f^*f) \text{ is dense in } E\}.$$

There is a bijective correspondence $t \mapsto f_t$, given by (10.1), (10.2), between $\mathcal{T}$ and $\mathcal{F}$, which is adjoint-preserving: $f_{t^*} = f_t^*$.

Proof. We already know (by (10.5)) that the mapping given by (10.1) takes $\mathcal{T}$ into $\mathcal{F}$.

Let $f \in \mathcal{F}$. Observe that $fp(f^*f) = p(ff^*)f$ for any polynomial p and hence by continuity for any p in $C([0,1])$. In particular,

$$f(1-f^*f)^{\frac{1}{2}} = (1-ff^*)^{\frac{1}{2}}f. \tag{10.6}$$

Define $D = \operatorname{ran}(1-f^*f)^{\frac{1}{2}}$. Then D is a dense submodule of E. Define $t: D \to F$ by $t(1-f^*f)^{\frac{1}{2}}x = fx$ $(x \in E)$. Notice that t is well-defined since $(1-f^*f)^{\frac{1}{2}}$ is injective. Thus we can write

$$t = f(1-f^*f)^{-\frac{1}{2}}. \tag{10.7}$$

By Lemma 10.3, $\operatorname{ran}(1-ff^*)$ is dense in F so we can similarly define $D^* = \operatorname{ran}(1-ff^*)^{\frac{1}{2}}$ and $s: D^* \to E$ by

$$s = f^*(1-ff^*)^{-\frac{1}{2}}. \tag{10.8}$$

Let $x = (1-f^*f)^{\frac{1}{2}}y \in D$, $z = (1-ff^*)^{\frac{1}{2}}w \in D^*$. Then

$$\begin{aligned}\langle tx, z\rangle &= \langle fy, (1-ff^*)^{\frac{1}{2}}w\rangle = \langle (1-ff^*)^{\frac{1}{2}}fy, w\rangle,\\ \langle x, sz\rangle &= \langle (1-f^*f)^{\frac{1}{2}}y, f^*w\rangle = \langle f(1-f^*f)^{\frac{1}{2}}y, w\rangle.\end{aligned}$$

Thus by (10.6) $z \in D(t^*)$ and $t^*z = sz$. So $t^* \supseteq s$ and in particular t^* is densely defined.

Let $q = (1-f^*f)^{\frac{1}{2}}$. From (10.8), together with (10.6) applied to f and to f^*, it follows that

$$s(fqu) = f^*fu \qquad (u \in E)$$

and

$$s(1-ff^*)v = f^*(1-ff^*)^{\frac{1}{2}}v = qf^*v \qquad (v \in F).$$

Therefore if we define $p \in \mathcal{L}(E \oplus F)$ by (9.7), p is a projection with $\operatorname{ran}(p) \subseteq G(t)$ and $\operatorname{ran}(1-p) \subseteq G(s) \subseteq G(t^*)$. This shows that t is closed, with $G(t) = \operatorname{ran}(p)$, and therefore regular by Proposition 9.5; and also that $s = t^*$, from which we deduce that $t_{f^*} = t_f^*$, where t_f denotes the regular operator constructed from f by (10.7).

It remains to show that the maps $t \mapsto f_t$ and $f \mapsto t_f$ are inverse to each other. One way round, this is easy: given t in $\mathcal{T}$, we know from (10.2) and (10.7) that $t = f_t(1-f_t^* f_t)^{-\frac{1}{2}} = t_{f_t}$. In the other direction, let $f \in \mathcal{F}$. Then $t_f = f(1-f^*f)^{-\frac{1}{2}}$ and (from the previous paragraph) $t_f^* = f^*(1-ff^*)^{-\frac{1}{2}}$. By (10.6) we have $(1-ff^*)^{-\frac{1}{2}}f(1-f^*f)^{\frac{1}{2}} = f$. Therefore, for all x in E,

$$\begin{aligned} f^*(1-ff^*)^{-\frac{1}{2}}f(1-f^*f)^{-\frac{1}{2}} \cdot (1-f^*f)x &= f^*fx, \\ t_f^* t_f(1-f^*f)x &= f^*fx, \\ (1+t_f^*t_f)(1-f^*f)x &= x. \end{aligned}$$

Hence $q_{t_f} = (1+t_f^*t_f)^{-\frac{1}{2}} = (1-f^*f)^{\frac{1}{2}}$ and (by (10.1))

$$f_{t_f} = t_f(1-f^*f)^{\frac{1}{2}} = f.$$

That concludes the proof of the theorem.

Notice that in the case $E = F$ it follows from Theorem 10.4 that t is selfadjoint if and only if f_t is selfadjoint.

If t is symmetric then, just as in the case of Hilbert space operators, we can investigate whether t has selfadjoint extensions by means of the Cayley transform. This will be our next main goal in this chapter. Let $t\colon E \to E$ be symmetric and regular, and write f, q for f_t, q_t. For x, y in E, we have

$$\begin{aligned} \langle tqx, qy \rangle &= \langle qx, tqy \rangle, \\ \langle fx, qy \rangle &= \langle qx, fy \rangle, \end{aligned}$$

and therefore

$$qf = f^*q; \tag{10.9}$$

and conversely if (10.9) holds then t is symmetric. Thus t is symmetric if and only if qf is selfadjoint.

Define w_+, w_- in $\mathcal{L}(E)$ by $w_\pm = f \pm iq = (t \pm i)q$. A simple calculation using (10.3) and (10.9) shows that $w_+^* w_+ = w_-^* w_- = 1$. Thus w_+, w_- are isometries and by Proposition 3.6 $\operatorname{ran}(w_\pm) = \operatorname{ran}(t \pm i)$ are complemented submodules of E. (This strengthens the conclusion of Lemma 9.7.)

Let $c_t = w_- w_+^* = (f - iq)(f^* - iq)$. Since w_+ and w_- are isometries, c_t is a partial isometry in $\mathcal{L}(E)$, with initial space $\operatorname{ran}(w_+) = \operatorname{ran}(t+i)$ and final space $\operatorname{ran}(w_-) = \operatorname{ran}(t-i)$. For x in E, we have

$$\begin{aligned}(1-c_t)c_t^* w_- x &= (1 - w_- w_+^*) w_+ x \\ &= (w_+ - w_-)x = 2iqx.\end{aligned} \tag{10.10}$$

Thus $\operatorname{ran}\big((1-c_t)c_t^*\big) \supseteq \operatorname{ran}(q)$, which is dense.

Putting $x = w_-^* y$ in (10.10), we see that

$$(1-c_t)c_t^* y = 2iqw_-^* y \qquad (y \in E),$$

so that $\operatorname{ran}\big((1-c_t)c_t^*\big) \subseteq \operatorname{ran}(q)$. Since we already have the reverse inclusion, it follows that $\operatorname{ran}\big((1-c_t)c_t^*\big) = \operatorname{ran}(q) = D(t)$. Also,

$$t(1-c_t)c_t^* y = 2itqw_-^* y = 2ifw_-^* y = i(w_+ + w_-)w_-^* y = i(1+c_t)c_t^* y.$$

Thus we can reconstruct t from c_t by the formulas

$$\begin{aligned} D(t) &= \operatorname{ran}\big((1-c_t)c_t^*\big), \\ t(1-c_t)c_t^* y &= i(1+c_t)c_t^* y \qquad (y \in E). \end{aligned} \tag{10.11}$$

If $x \in D(t)$ then $x = qy$ for some y in E, and

$$\begin{aligned} c_t(t+i)x = c_t(t+i)qy &= c_t(f+iq)y \\ &= w_- w_+^* w_+ y \\ &= w_- y = (f - iq)y = (t-i)x. \end{aligned}$$

This shows that the partial isometry c_t is specified on its initial space $\operatorname{ran}(t+i)$ by the formula

$$c_t(t+i)x = (t-i)x \qquad (x \in D(t)), \tag{10.12}$$

which justifies the use of the term *Cayley transform* for the operator c_t.

Denote by $\mathcal{C}$ the set of all partial isometries $c \in \mathcal{L}(E)$ such that $(1-c)c^*$ has dense range. If c_1, c_2 are partial isometries in $\mathcal{L}(E)$ then we say that c_2 is an extension of c_1 if the initial space of c_2 contains the initial space F of c_1 and $c_2 x = c_1 x$ for all x in F. It is clear that if $c_1 \in \mathcal{C}$ and c_2 is an extension of c_1 then $c_2 \in \mathcal{C}$.

THEOREM 10.5. *Let E be a Hilbert A-module, let $\mathcal{S}$ denote the set of symmetric regular operators on E and let $\mathcal{C}$ be as above. There is a bijective correspondence $t \mapsto c_t$ between $\mathcal{S}$ and $\mathcal{C}$ in which s is an extension of t if and only if c_s is an extension of c_t.*

Proof. We have already shown that the map $t \mapsto c_t$ takes $\mathcal{S}$ into $\mathcal{C}$. Suppose that $c \in \mathcal{C}$. Define t by $D(t) = \operatorname{ran}\big((1-c)c^*\big)$ and

$$t \cdot (1-c)c^* x = i(1+c)c^* x. \tag{10.13}$$

Then $\operatorname{ran}\big((1-c)c^*c\big) = \operatorname{ran}\big((1-c^*)c\big) \subseteq D(t)$, and

$$\begin{aligned} t \cdot (1-c^*)cx &= -t(1-c)c^*cx \\ &= -i(1+c)c^*cx \\ &= -i(1+c^*)cx. \end{aligned} \tag{10.14}$$

Note that t is closed: for if (x_n) is a sequence in E with $(1-c)c^*x_n \to y$ and $i(1+c)c^*x_n \to z$ (for some y, z in E) then $c^*x_n \to \frac{1}{2}(y - iz)$. Since $\operatorname{ran}(c^*)$ is closed, we must have $\frac{1}{2}(y - iz) = c^*x$ for some x in E, from which it follows that $y = (1-c)c^*x \in D(t)$ and $ty = z$.

Next, observe that $\operatorname{ran}(1-c^*) \subseteq D(t^*)$: for if $x = (1-c)c^*x_0 \in D(t)$ and $y = (1-c^*)z$ then by (10.13)

$$\begin{aligned} \langle tx, y\rangle &= \langle i(1+c)c^*x_0, (1-c^*)z\rangle \\ &= -i\langle (1-c^2)c^*x_0, z\rangle \\ &= -i\langle (1+c)x, z\rangle \\ &= \langle x, -i(1+c^*)z\rangle. \end{aligned}$$

This shows that $y \in D(t^*)$ and

$$t^*(1-c^*)z = -i(1+c^*)z \qquad (z \in E). \tag{10.15}$$

A similar argument using (10.14) in place of (10.13) shows that we have $\operatorname{ran}(1-c) \subseteq D(t^*)$, with

$$t^*(1-c)z = i(1+c)z \qquad (z \in E). \tag{10.16}$$

Comparison of (10.16) with (10.13) shows that $t \subseteq t^*$, so that t is symmetric.

For x in E, let $y = (1-c)c^*x + (1-c^*)cx$. By (10.13) and (10.14), $y \in D(t)$ and

$$\begin{aligned} ty &= i(1+c)c^*x - i(1+c^*)cx \\ &= i[(1-c)(2-c^*)x - (1-c^*)(2-c)x]. \end{aligned}$$

It follows from (10.16) and (10.15) that $ty \in D(t^*)$ and that

$$\begin{aligned} t^*ty &= i\big[i(1+c)(2-c^*)x + i(1+c^*)(2-c)x\big] \\ &= -4x - y. \end{aligned}$$

Therefore $x = (1+t^*t)(-\frac{1}{4}y) \in \operatorname{ran}(1+t^*t)$, so that $1+t^*t$ is surjective and hence t is regular.

If we denote by t_c the symmetric regular operator constructed from c in this way then it follows from (10.11) and (10.13) that $t = t_{c_t}$. On the other hand, suppose that $c \in \mathcal{C}$ and $x \in E$. Then $(1-c)c^*x \in D(t_c)$, and by (10.13)

$$(t_c + i)(1-c)c^*x = i(1+c)c^*x + i(1-c)c^*x = 2ic^*x. \tag{10.17}$$

Thus $\operatorname{ran}(t_c + i) = \operatorname{ran}(c^*)$, which is the initial space of c. A similar calculation shows that

$$(t_c - i)(1-c)c^*x = 2icc^*x. \tag{10.18}$$

According to (10.12), c_{t_c} is the partial isometry with initial space $\operatorname{ran}(t_c+i)$ which takes $(t_c+i)z$ to $(t_c-i)z$ $\;(z \in D(t_c))$. From (10.17) and (10.18) we see that this implies that $c_{t_c} = c$.

Thus the maps $t \mapsto c_t$ and $c \mapsto t_c$ are inverse to each other. Finally, it is clear from the description of these maps given by (10.11) and (10.12) that they preserve the respective notions of extension in $\mathcal{S}$ and $\mathcal{C}$.

PROPOSITION 10.6. *Let E be a Hilbert A-module and let t be a densely defined symmetric operator on E. Then t is regular if and only if the submodules* $\operatorname{ran}(t \pm i)$ *are complemented in E.*

Proof. We have already observed the implication in one direction (see the paragraph between (10.9) and (10.10)).

Suppose that t is densely defined and symmetric, and that $\operatorname{ran}(t \pm i)$ are complemented. In particular, $\operatorname{ran}(t \pm i)$ are closed, from which it easily follows that t must be closed. Write $e_\pm$ for the projections onto $\operatorname{ran}(t \pm i)$. For x in E, let

$$\begin{aligned} s_+x &= e_-(t+i)^{-1}e_+x, \\ s_-x &= e_+(t-i)^{-1}e_-x. \end{aligned}$$

Then $s_\pm$ are contractions by the proof of Lemma 9.7.

Given x, y in E, we can write $e_+x = (t+i)x_0$, $e_-y = (t-i)y_0$, where $x_0, y_0 \in D(t)$. Then

$$\begin{aligned}\langle s_+x, y\rangle = \langle e_-x_0, y\rangle &= \langle x_0, e_-y\rangle \\ &= \langle x_0, (t-i)y_0\rangle \\ &= \langle (t+i)x_0, y_0\rangle \\ &= \langle e_+x, y_0\rangle = \langle x, e_+y_0\rangle = \langle x, s_-y\rangle.\end{aligned}$$

Thus the operators s_+, s_- are adjointable, and are in fact each other's adjoints in $\mathcal{L}(E)$.

Let $c = e_-e_+ - 2is_+ \in \mathcal{L}(E)$. Then

$$c(t+i)x = e_-(t+i)x - 2ie_-x = e_-(t-i)x = (t-i)x \qquad (x \in D(t)). \tag{10.19}$$

By the proof of Lemma 9.7, c is isometric on $\operatorname{ran}(e_+)$; and clearly $c = 0$ on $\operatorname{ran}(e_+)^\perp$. Thus c is a partial isometry.

If $x \in D(t)$ then

$$(1-c)c^*(t-i)x = (1-c)(t+i)x = (t+i)x - (t-i)x = 2ix, \tag{10.20}$$

so that $\operatorname{ran}\big((1-c)c^*\big) \supseteq D(T)$, which is dense. Therefore $c \in \mathcal{C}$. Conversely, if $(1-c)c^*y \in \operatorname{ran}\big((1-c)c^*\big)$ then we may assume that y is in $\operatorname{ran}(cc^*) = \operatorname{ran}(t-i)$, say $y = (t-i)y_0$ where $y_0 \in D(t)$. From this it is easy to establish, using (10.11), that $t = t_c$, and therefore t is regular.

It follows from Lemma 9.8 that a symmetric regular operator t on E is selfadjoint if and only if c_t is unitary. Thus the problem of whether a symmetric regular operator has a selfadjoint extension is equivalent to that of finding a unitary extension of its Cayley transform. If the C*-algebra $\mathcal{L}(E)$ satisfies a suitable finiteness condition then this problem can become trivial. For example, if $\mathcal{L}(E)$ is finite in the sense of having no non-unitary isometries then every symmetric regular operator on E must automatically be selfadjoint (since $c_t = w_-w_+^*$ and the operators $w_\pm$ are isometries).

At the end of the previous chapter, we investigated regular operators on interior tensor products of Hilbert C*-modules. We shall now use the bounded transform to look at regular operators on exterior tensor products.

Suppose that E, E' are Hilbert A-modules, F, F' are Hilbert B-modules and $t_1: E \to E'$, $t_2: F \to F'$ are regular. We want to define $t_1 \otimes t_2$ as a regular operator from $E \otimes F$ to $E' \otimes F'$, with $(t_1 \otimes t_2)^* = t_1^* \otimes t_2^*$. Even in the case of Hilbert space operators, this is not a straightforward construction, and in fact it is tantamount to a proof of the commutation theorem for von Neumann algebras (see Section 11.2 of [KadRin]). The following argument, when specialised to the case of Hilbert space operators, yields a very efficient approach to the construction of the tensor product of closable operators, which illustrates clearly the power of the bounded transform as a technique for handling closed operators.

Starting at the algebraic level, let

$$D_0 = D(t_1) \otimes_{\mathrm{alg}} D(t_2), \quad D_0^* = D(t_1^*) \otimes_{\mathrm{alg}} D(t_2^*).$$

These are dense submodules of $E \otimes F$, $E' \otimes F'$ respectively. For x in $D(t_1)$ and y in $D(t_2)$, define $t_1 \otimes_{\mathrm{alg}} t_2 \cdot x \otimes y = t_1 x \otimes t_2 y$. This extends by linearity to define $t_1 \otimes_{\mathrm{alg}} t_2$ as an operator from D_0 to $E' \otimes F'$. It is easy to check that $t_1^* \otimes_{\mathrm{alg}} t_2^* \subseteq (t_1 \otimes_{\mathrm{alg}} t_2)^*$. Therefore $(t_1 \otimes_{\mathrm{alg}} t_2)^*$ is densely defined and so $t_1 \otimes_{\mathrm{alg}} t_2$ is closable. We define $t_1 \otimes t_2$ to be its closure.

For $i = 1, 2$ let f_i, q_i be the bounded operators associated with t_i as in (10.1), (10.3). By Theorem 10.4 the corresponding operators associated with t_i^* are f_i^* and r_i, where $r_i = (1 - f_i f_i^*)^{\frac{1}{2}}$. Note that by (10.6) we have

$$f_i q_i = r_i f_i. \tag{10.21}$$

Define s in $\mathcal{L}\big((E \otimes F) \oplus (E' \otimes F')\big)$ by

$$s = \begin{pmatrix} q_1 \otimes q_2 & -f_1^* \otimes f_2^* \\ f_1 \otimes f_2 & r_1 \otimes r_2 \end{pmatrix}.$$

Using (10.21), we find

$$ss^* = \begin{pmatrix} q_1^2 \otimes q_2^2 + f_1^* f_1 \otimes f_2^* f_2 & 0 \\ 0 & f_1 f_1^* \otimes f_2 f_2^* + r_1^2 \otimes r_2^2 \end{pmatrix}.$$

Since $\mathrm{ran}(q_1^2 \otimes q_2^2)$ is dense in $E \otimes F$, it follows by Corollary 10.2 that so is $\mathrm{ran}(q_1^2 \otimes q_2^2 + f_1^* f_1 \otimes f_2^* f_2)$, and similarly for the other diagonal element of ss^*. Therefore $\mathrm{ran}(ss^*)$ is dense and hence by Proposition 3.7 so is $\mathrm{ran}(s)$. But if $x, y \in E \otimes F$ then

$$s\begin{pmatrix} x \\ y \end{pmatrix} = \begin{pmatrix} q_1 \otimes q_2 \cdot x \\ f_1 \otimes f_2 \cdot x \end{pmatrix} + \begin{pmatrix} -f_1^* \otimes f_2^* \cdot y \\ r_1 \otimes r_2 \cdot y \end{pmatrix}. \tag{10.22}$$

The first term on the right-hand side of (10.22) is in $G(t_1 \otimes t_2)$ and the second term is in $vG(t_1^* \otimes t_2^*)$ (where v has the same meaning as in Chapter 9). Thus $G(t_1 \otimes t_2) \oplus vG(t_1^* \otimes t_2^*)$ is dense in, and is therefore the whole of, $(E \otimes F) \oplus (E' \otimes F')$. It follows from Proposition 9.5 that $t_1 \otimes t_2$ is regular, and furthermore that $(t_1 \otimes t_2)^* = t_1^* \otimes t_2^*$, as required.

Our last major topic in this chapter is that of unbounded multipliers. An unbounded multiplier on a C*-algebra A is by definition a regular mapping on the Hilbert A-module A. Specifically, an unbounded multiplier on A is a closed A-linear map $m: J \to A$, where J is a dense right ideal in A, with a densely defined adjoint m^*, such that $1+m^*m$ has dense range. We denote by $UM(A)$ the set of all unbounded multipliers on A. Note that, although we call these operators unbounded, the definition encompasses bounded multipliers also, so that $M(A) \subseteq UM(A)$. If A is unital then A has no proper dense right ideals and so $M(A) = UM(A)$. Note also that the set $UM(A)$ has very little algebraic structure. It has an adjoint operation and is closed under scalar multiplication, but not in general under addition or multiplication.

If X is a locally compact Hausdorff space then $UM(C_0(X))$ can be identified with $C(X)$, the $*$-algebra of all continuous complex-valued functions on X. This is an easy consequence of the fact that $m = z_m(1 - z_m^* z_m)^{-\frac{1}{2}}$, where z_m is the bounded transform of m in $M(C_0(X)) \cong C_b(X)$. Thus when A is commutative, $UM(A)$ does in fact have a good algebraic structure.

If $A = K$ (the algebra of compact operators on the Hilbert space H) then $UM(A)$ can be identified with the set of densely defined closed operators on H. The reason (similar to that in the previous paragraph) is as follows. If $m \in UM(K)$ then $m = z_m(1 - z_m^* z_m)^{-\frac{1}{2}}$, where this time $z_m \in M(K) \cong \mathcal{L}(H)$. From (10.5), z_m, regarded as an element of $\mathcal{L}(H)$, is a contraction such that $1 - z_m^* z_m$ has closed range. So by Theorem 10.4, applied to the Hilbert $\mathbb{C}$-module H, z_m is the bounded transform of a closed operator on H.

Let A, B be C*-algebras, let E be a Hilbert A-module and suppose that $\alpha: B \to \mathcal{L}(E)$ is a nondegenerate $*$-homomorphism. We shall write $m \mapsto z_m$ for the bounded transform on $UM(B)$ and $t \mapsto f_t$ for the bounded transform on regular operators on E.

PROPOSITION 10.7. *With $\alpha\colon B \to \mathcal{L}(E)$ as above and $m \in UM(B)$, there is a regular operator $\hat{\alpha}(m)$ on E such that if $b \in D(m)$ then*

$$\operatorname{ran}\big(\alpha(b)\big) \subseteq D\big(\hat{\alpha}(m)\big), \qquad \hat{\alpha}(m)\alpha(b) = \alpha(mb).$$

Also, $\alpha(z_m) = f_{\hat{\alpha}(m)}$; and $\hat{\alpha}(m^*) = \hat{\alpha}(m)^*$.

Proof. Let $\alpha_*(m)$ be the regular operator on $B \otimes_\alpha E$ constructed in Proposition 9.10, so that $\alpha_*(m^*) = \alpha_*(m)^*$ and

$$\alpha_*(m) \cdot b \dot{\otimes} x = mb \dot{\otimes} x \qquad (b \in D(m),\ x \in E).$$

Making use of the canonical unitary equivalence $b \dot{\otimes} x \mapsto \alpha(b)x$ between $B \otimes_\alpha E$ and E (see Chapter 4), we obtain a regular operator $\hat{\alpha}(m)$ on E satisfying $\hat{\alpha}(m^*) = \hat{\alpha}(m)^*$ and

$$\hat{\alpha}(m)\alpha(b)x = \alpha(mb)x \qquad (b \in D(m),\ x \in E). \tag{10.23}$$

It remains to show that $\alpha(z_m) = f_{\hat{\alpha}(m)}$. For this, let

$$q_1 = (1 + m^*m)^{-\frac{1}{2}}, \qquad q_2 = \big(1 + \hat{\alpha}(m)^*\hat{\alpha}(m)\big)^{-\frac{1}{2}}.$$

Since $(1 + m^*m)q_1^2 = 1$, it follows from (10.23) (applied to m^* and then to m) that $\big(1+\hat{\alpha}(m)^*\hat{\alpha}(m)\big)\alpha(q_1)^2 = 1$, from which we deduce that $q_2 = \alpha(q_1)$. Then (by (10.1) and (10.23))

$$\alpha(z_m) = \alpha(mq_1) = \hat{\alpha}(m)q_2 = f_{\hat{\alpha}(m)}.$$

That concludes the proof of the proposition.

We aim to set up a functional calculus for selfadjoint regular operators. As a preliminary to this, we need the following technical result.

LEMMA 10.8. *With α, m as in Proposition 10.7, if D_0 is a core for m then $\alpha(D_0)E$ is a core for $\hat{\alpha}(m)$.*

Proof. In the notation used in the proof of Proposition 10.7, $\alpha_*(m)$ is the closure of the operator $m \dot{\otimes} \iota$ on the image of $D_0 \otimes_{\text{alg}} E$ in $B \otimes_\alpha E$. Transferring this information to E by means of the canonical unitary equivalence $B \otimes_\alpha E \approx E$, we obtain the result of the lemma.

THEOREM 10.9. *Let E be a Hilbert A-module and let t be a selfadjoint regular operator on E. There is a $*$-homomorphism $f \mapsto f(t)$ from $C(\mathbf{R})$ into the regular operators on E such that $\iota(t) = t$ and $\theta(t) = f_t$, where ι is the function $\lambda \mapsto \lambda$ and θ is the function $\lambda \mapsto \lambda(1+\lambda^2)^{-\frac{1}{2}}$ $(\lambda \in \mathbf{R})$.*

Proof. The map θ is a homeomorphism from $\mathbf{R}$ onto $(-1,1)$ (with inverse given by $\theta^{-1}(\mu) = \mu(1-\mu^2)^{-\frac{1}{2}}$, $-1 < \mu < 1$).

By Theorem 10.4 the bounded transform f_t of t is selfadjoint, so we can define a $*$-homomorphism $\alpha: C_0(\mathbf{R}) \to \mathcal{L}(E)$ by

$$\alpha(g) = (g \circ \theta^{-1})(f_t) \qquad (g \in C_0(\mathbf{R})).$$

If g is the function $\lambda \mapsto (1+\lambda^2)^{-1}$ then $g \circ \theta^{-1}(\mu) = 1 - \mu^2$ and so

$$\alpha(g) = 1 - f_t^2 = q_t^2,$$

which has dense range. By Lemma 10.1, $\alpha(g^{\frac{1}{n}}) = \alpha(g)^{\frac{1}{n}} \to 1$ strictly in $\mathcal{L}(E)$. But $(g^{\frac{1}{n}})$ is an approximate unit for $C_0(\mathbf{R})$, so by Proposition 2.5 α is nondegenerate.

Thus by Proposition 10.7, for every f in $C(\mathbf{R}) \cong UM(C_0(\mathbf{R}))$ we have a regular operator $\hat{\alpha}(f)$ on E. We define $f(t) = \hat{\alpha}(f)$. By Proposition 10.7 the map $f \mapsto f(t)$ is adjoint-preserving. If $f, g \in C(\mathbf{R})$ and $h \in C_{00}(\mathbf{R})$ then h is in the domains of $f+g$ and fg (regarded as unbounded multipliers of $C_0(\mathbf{R})$), and by Proposition 10.7

$$\hat{\alpha}(f+g)\alpha(h) = \big(\hat{\alpha}(f) + \hat{\alpha}(g)\big)\alpha(h),$$
$$\hat{\alpha}(fg)\alpha(h) = \hat{\alpha}(f)\hat{\alpha}(g)\alpha(h).$$

But $C_{00}(\mathbf{R})$ is a core for every element of $C(\mathbf{R})$, so we see from Lemma 10.8 that $\hat{\alpha}(f+g) = \hat{\alpha}(f)+\hat{\alpha}(g)$ and $\hat{\alpha}(fg) = \hat{\alpha}(f)\hat{\alpha}(g)$. Thus the map $f \mapsto f(t)$ is a $*$-homomorphism.

It is evident from the definition of α that $\theta(t) = f_t$. If q is the function $\lambda \mapsto (1+\lambda^2)^{-\frac{1}{2}}$ then $\alpha(q) = (1-f_t^2)^{\frac{1}{2}} = q_t$. Since $\theta = \iota q$, we have $q \in D(\iota)$ and by Proposition 10.7 $f_t = \hat{\alpha}(\iota)q_t$. Hence $\hat{\alpha}(\iota) \supseteq t$. But ι and t are both selfadjoint and therefore $\hat{\alpha}(\iota) = t$.

It is shown in [Wor 5] and [WorNap] how the functional calculus for selfadjoint regular operators given by Theorem 10.9 can also be defined for normal regular operators, where normality for regular operators is defined in such a way that t is normal if and only if f_t is a normal element of $\mathcal{L}(E)$.

References for Chapter 10: [Wor 5], [WorNap]; the definition of an unbounded multiplier was first given in [Con].

Chapter 11

What next?

The material in the previous three chapters has been motivated by the desire to develop a C*-algebraic theory of quantum groups. Recall from Chapter 8 that a C*-bialgebra is a C*-algebra A together with a comultiplication $\delta \in \mathrm{Mor}(A, A \otimes A)$ satisfying the two conditions on page 82. In the case where A is commutative, say $A = C_0(G)$, a comultiplication is specified by an associative multiplication on G, by the formula (8.5). The astute reader will have observed that for this formula to make sense it is not necessary for G to be a group—it would suffice for G to be a (locally compact) semigroup. Indeed, a C*-bialgebra is best thought of as a "quantum semigroup". For the algebra $C_0(G)$ to reflect the fact that the group G has an identity and an inverse operation, some additional structure is called for.

In the algebraic theory of Hopf algebras, it is well known how to achieve this: a Hopf algebra is a bialgebra (A, δ) (over $\mathbb{C}$, say) together with two additional operations, a *counit*, which is a homomorphism $\varepsilon: A \to \mathbb{C}$, and an *antipode*, which is an antihomomorphism $\kappa: A \to A$, satisfying certain natural conditions. These maps reflect at the algebra level the existence of an identity element and an inverse operation in the group. In the case where $A = C_0(G)$ for a group G with identity e, they are given by

$$\varepsilon(f) = f(e), \quad \kappa(f)(s) = f(s^{-1}) \qquad (f \in C_0(G),\ s \in G).$$

For the group convolution C*-algebra $C_r^*(G)$, the maps ε, κ are given on $C_{00}(G)$ by

$$\varepsilon(f) = \int_G f, \quad \kappa(f)(s) = \Delta(s)f(s^{-1}) \qquad (f \in C_{00}(G),\ s \in G).$$

It is at this point that difficulties begin to show themselves. In the C*-algebraic theory of quantum groups, the mappings ε, κ cannot in general be defined as bounded mappings on the whole algebra. Therefore one would expect a C*-Hopf algebra to be a C*-bialgebra (A, δ) together with (possibly unbounded) densely defined maps ε, κ with domain in A and range in $\mathbb{C}$, A respectively, satisfying certain axioms. These axioms should of course hold in the case of group function algebras and convolution algebras, and also for the many examples of "nonclassical" C*-algebraic quantum groups that are now known ([EffRua], [PodWor], [Rie 3], [Rie 4], [Wor 2], [Wor 4], [Wor 5], [Wor 6]). They should be reasonably straightforward to verify in particular examples, and they should be strong enough to support a theory which is sufficiently rich to include such results as the existence and uniqueness of a Haar weight and a duality theory extending the Pontryagin duality for locally compact abelian groups.

To date, an axiomatic basis for C*-algebraic quantum group theory has not been established, though it seems very likely that such a theory should be achievable. Much progress has already been made: the compact case ([Wor 3], [Wor 7]) and the discrete case ([EffRua], [Van]) are now well understood. Powerful tools for proving duality results are established in [BaaSka].

The reader who is industrious enough to absorb all the material in the references cited in this chapter will be well placed to follow (or indeed to take a lead in) future developments in this rich and intriguing field of research.

References

[AkePedTom]
C.A. AKEMANN, G.K. PEDERSEN and J. TOMIYAMA, Multipliers of C*-algebras, *J. Functional Anal.* 13 (1973) 277–301.

[BaaJul]
S. BAAJ and P. JULG, Théorie bivariante de Kasparov et opérateurs non bornés dans les C*-modules hilbertiens, *C.R. Acad. Sci. Paris* 296 (1983) 875–878.

[BaaSka]
S. BAAJ and G. SKANDALIS, Unitaires multiplicatifs et dualité pour les produits croisés de C*-algèbres, *Ann. Sci. École Norm. Sup.* (4) 26 (1993) 425–488.

[Bla]
B. BLACKADAR, *K-theory for operator algebras* (MSRI Publ. 5), Springer-Verlag, 1986.

[Bro]
L.G. BROWN, Stable isomorphism of hereditary subalgebras of C*-algebras, *Pacific J. Math.* 71 (1977) 335–348.

[BroGreRie]
L.G. BROWN, P. GREEN and M.A. RIEFFEL, Stable isomorphism and strong Morita equivalence of C*-algebras, *Pacific J. Math.* 71 (1977) 349–363.

[Bus]
R.C. BUSBY, Double centralizers and extensions of C*-algebras, *Trans. Amer. Math. Soc.* 132 (1968) 79–99.

[Con]
A. CONNES, An analogue of the Thom isomorphism for crossed products of a C*-algebra by an action of **R**, *Adv. in Math.* 39 (1981) 31–55.

[CunHig]
J. CUNTZ and N. HIGSON, Kuiper's theorem for Hilbert modules, in *Operator algebras and mathematical physics* (ed. P.E.T. Jørgensen and P.S. Muhly), Contemp. Math. 62 (Amer. Math. Soc., 1987), pp. 429–434.

[Dix 1]
J. DIXMIER, *Les algèbres d'opérateurs dans l'espace hilbertien*, 2nd edition, Gauthier-Villars, 1969.

[Dix 2]
J. DIXMIER, *Les C*-algèbres et leurs représentations*, Gauthier-Villars, 1964.

[EffRua]
E.G. EFFROS and Z.-J. RUAN, Discrete quantum groups, I. The Haar measure, preprint, UCLA, 1993.

[Fil]
O.G. FILIPPOV, On C*-algebras A over which the Hilbert module $l_2(A)$ is selfdual, *Vestnik Moskov. Univ. Ser. I Mat. Mekh.* 1987, no. 4, 74–76 (Russian); *Moscow Univ. Math. Bull.* 42 (1987) 87–90 (English).

[Fra]
M. FRANK, Self-duality and C*-reflexivity of Hilbert C*-moduli, *Z. Anal. Anwendungen* 9 (1990) 165–176.

[HewRos]
E. HEWITT and K.A. ROSS, *Abstract harmonic analysis*, Vol. I, Springer-Verlag, 1963.

[Hil]
M. HILSUM, Fonctorialité en K-théorie bivariante pour les variétés lipschitziennes, *K-theory* 3 (1989) 401–440.

[Ior]
V. DE M. IORIO, Hopf C*-algebras and locally compact groups, *Pacific J. Math.* 87 (1980) 75–96.

[JenTho]
K.K. JENSEN and K. THOMSEN, *Elements of KK-theory*, Birkhäuser, 1991.

[KadRin]
R.V. KADISON and J.R. RINGROSE, *Fundamentals of the theory of operator algebras*, Vols. I, II, Academic Press, 1983, 1986.

[Kap]
I. KAPLANSKY, Modules over operator algebras, *Amer. J. Math.* 75 (1953) 839–853.

[Kas 1]
G.G. KASPAROV, Hilbert C*-modules: Theorems of Stinespring and Voiculescu, *J. Operator Theory* 4 (1980) 133–150.

[Kas 2]
G.G. KASPAROV, The operator K-functor and extensions of C*-algebras, *Izv. Akad. Nauk SSSR Ser. Mat.* 44 (1980) no. 3, 571–636 (Russian); *Math. USSR–Izv.* 16 (1981) no. 3, 513–572 (English).

[Lan 1]
E.C. LANCE, Unitary operators on Hilbert C*-modules, *Bull. London Math. Soc.*, to appear.

[Lan 2]
M.B. LANDSTAD, Duality theory of covariant systems, *Trans. Amer. Math. Soc.* 248 (1979) 223–267.

[Min]
J.A. MINGO, K-theory and multipliers of stable C*-algebras, *Trans. Amer. Math. Soc.* 299 (1987) 397–411.

[MinPhi]
J.A. MINGO and W.J. PHILLIPS, Equivariant triviality theorems for Hilbert C*-modules, *Proc. Amer. Math. Soc.* 91 (1984) 225–230.

[Miš]
A.S. MIŠČENKO, Banach algebras, pseudodifferential operators and their applications in K-theory, *Uspekhi Mat. Nauk* 34 (1979) no. 6, 67–79 (Russian); *Russian Math. Surveys* 34 (1979) no. 6, 77–91 (English).

[Mur]
G.J. MURPHY, *C*-algebras and operator theory*, Academic Press, 1990.

[Neu]
J. von Neumann, On a certain topology for rings of operators, *Ann. of Math.* 37 (1936) 111–115.

[Pas]
W.L. Paschke, Inner product modules over B*-algebras, *Trans. Amer. Math. Soc.* 182 (1973) 443–468.

[Pau]
V.I. Paulsen, *Completely bounded maps and dilations* (Pitman Research Notes in Math. Ser. 146), Longman, 1986.

[Ped]
G.K. Pedersen, *C*-algebras and their automorphism groups* (London Math. Soc. Monographs 14), Academic Press, 1979.

[Pie]
R.S. Pierce, *Associative algebras* (Graduate Texts in Math. 88), Springer-Verlag, 1982.

[PodWor]
P. Podleś and S.L. Woronowicz, Quantum deformation of Lorentz group, *Commun. Math. Phys.* 130 (1990) 381–431.

[Rie 1]
M.A. Rieffel, Induced representations of C*-algebras, *Adv. in Math.* 13 (1974) 176–257.

[Rie 2]
M.A. Rieffel, Morita equivalence for operator algebras, in *Operator algebras and applications* (ed. R.V. Kadison), Proc. Sympos. Pure Math. Vol. 38, Part I (Amer. Math. Soc., 1982), pp. 285–298.

[Rie 3]
M.A. Rieffel, Some solvable quantum groups, in *Operator algebras and topology (Craiova, 1989)*, Pitman Res. Notes Math. Ser. 270 (Longman, 1992), pp. 146–159.

[Rie 4]
M.A. Rieffel, *Deformation quantization for actions of* $\mathbf{R}^d$ (Mem. Amer. Math. Soc. 506 (Vol. 106, no. 1)), Amer. Math. Soc., 1993.

[Rud]
W. RUDIN, *Functional analysis*, Tata McGraw-Hill, 1973.

[Tak]
M. TAKESAKI, *Theory of operator algebras I*, Springer-Verlag, 1979.

[Tay]
D.C. TAYLOR, The strict topology for double centralizer algebras, *Trans. Amer. Math. Soc.* 150 (1970) 633–644.

[Val]
J.-M. VALLIN, C*-algèbres de Hopf et C*-algèbres de Kac, *Proc. London Math. Soc.* (3) 50 (1985) 131–174.

[Van]
A. VAN DAELE, Discrete quantum groups, preprint, Leuven, 1993.

[Weg]
N.E. WEGGE-OLSEN, *K-theory and C*-algebras*, Oxford University Press, 1993.

[Wor 1]
S.L. WORONOWICZ, Pseudospaces, pseudogroups, and Pontryagin duality, in *Mathematical problems in theoretical physics* (Proc. Internat. Conf. Math. Phys., Lausanne, 1979, ed. K. Osterwalder), Lecture Notes in Phys. 116 (Springer-Verlag, 1980), pp. 407–412.

[Wor 2]
S.L. WORONOWICZ, Twisted SU(2) group. An example of a non-commutative differential calculus, *Publ. RIMS Kyoto* 23 (1987) 117–181.

[Wor 3]
S.L. WORONOWICZ, Compact matrix pseudogroups, *Commun. Math. Phys.* 111 (1987) 613–665.

[Wor 4]
S.L. WORONOWICZ, Tannaka–Krein duality for compact matrix pseudogroups. Twisted SU(N) groups, *Invent. Math.* 93 (1988) 35–76.

[Wor 5]
S.L. WORONOWICZ, Unbounded elements affiliated with C*-algebras and non-compact quantum groups, *Commun. Math. Phys.* 136 (1991) 399–432.

[Wor 6]
S.L. Woronowicz, Quantum E(2)-group and its Pontryagin dual, *Lett. Math. Phys.* 23 (1991) 251–263.

[Wor 7]
S.L. Woronowicz, Compact quantum groups, preprint, Warsaw, 1992.

[WorNap]
S.L. Woronowicz and K. Napiórkowski, Operator theory in the C*-algebra framework, *Rep. Math. Phys.* 31 (1992) 353–371.

Index

absorption, 60.
adjointable map, 8.

bounded transform, 107.

C*-bialgebra, 82, 121.
Cauchy–Schwarz inequality, 3.
Cayley transform, 112.
"compact operator", 10.
complemented submodule, 21.
completely positive map, 39, 45–47, 78.
completion, 4.
comultiplication, 82.
conditional expectation, 7.
core, 97.
corepresentation
  unbounded, 104–106.
  unitary, 87.

densely defined operator
  non-regular, 103.
  regular, 96, 109–111.
  selfadjoint, 101.
  symmetric, 101, 113.
direct sum, 5–6.

essential ideal, 14.
exterior tensor product, 34–38, 62, 116.

"finite-rank" operator, 9.
fully complemented submodule, 61.
function C*-bialgebra, 83, 122.
functional calculus for selfadjoint regular operators, 118–120.

Hilbert C*-module, 4.
  countably generated, 60, 66.
  full, 69.

idealiser, 17.
inner-product C*-module, 2.
interior tensor product, 38–44, 105.
isometry, 26.

KSGNS construction, 50–52.

leg notation, 80.
localisation, 7, 57.

Morita equivalence, 74.
morphism
  of C*-algebras, 18.
  of C*-bialgebras, 82.
multiplier algebra, 14–18.

nondegenerate
  *-homomorphism, 15, 19.
  completely positive map, 48–49, 53.

partial isometry, 29–30, 112–115.
polar decomposition, 29.
product of sets, 15.

quantum group, iv, 121–122.

reduced convolution C*-bialgebra,
    83–86, 90–93, 122.
regular operator, 96.
retraction, 55.
Riesz–Fréchet theorem, 13.

self-dual Hilbert C*-module, 13.
semi-inner-product C*-module, 3.
σ-unital C*-algebra, 60, 66.
slice map, 79.
stabilisation, 60.
stable
    algebra, 63, 69.
    multiplier algebra, 63.
    isomorphism, 69, 74.
Stone–Čech compactification, 15.
strict
    completely positive map, 49.
    topology, 11, 18, 76–77.
strictly positive, 59.
    mod A, 108.

tensor product
    exterior, 34–38, 62, 116.
    interior, 38–44, 105.
    of C*-algebras, 31.
    of Hilbert spaces, 31.
Tietze theorem for C*-algebras, 66.

unbounded multiplier, 117.
    functional calculus, 118.
unitary
    adjointable operator, 24.
    corepresentation, 87–88.
    equivalence, 24.

Printed in the United States
By Bookmasters